# User-Innovation

# Routledge Studies in Innovation, Organization and Technology

**1. Innovation in the U.S. Service Sector**
Michael P. Gallaher, Albert N. Link and Jeffrey E. Petrusa

**2. Information and Communications Technologies in Society**
E-Living in a Digital Europe
Edited by Ben Anderson, Malcolm Brynin and Yoel Raban

**3. The Innovative Bureaucracy**
Bureaucracy in an Age of Fluidity
Alexander Styhre

**4. Innovations and Institutions**
An Institutional Perspective on the Innovative Efforts of Banks and Insurance Companies
Patrick Vermeulen and Jorg Raab

**5. Knowledge and Innovation in Business and Industry**
The Importance of Using Others
Edited by Håkan Håkansson & Alexandra Waluszewski

**6. Knowledge and Innovation**
A Comparative Study of the USA, the UK and Japan
Helen Brown

**7. Industrial Innovation in Japan**
Edited by Takuji Hara, Norio Kambayashi and Noboru Matsushima

**8. Managing and Marketing Radical Innovations**
Marketing New Technology
Birgitta Sandberg

**9. Mobility and Technology in the Workplace**
Edited by Donald Hislop

**10. Energizing Management Through Innovation and Entrepreneurship**
European Research and Practice
Edited by Milé Terziovski

**11. Innovating for Sustainability**
Green Entrepreneurship in Personal Mobility
Luca Berchicci

**12. Organizational Capital**
Modelling, Measuring and Contextualising
Edited by Ahmed Bounfour

**13. User-Innovation**
Barriers to Democratization and IP Licensing
Viktor Braun and Cornelius Herstatt

# User-Innovation

## Barriers to Democratization and IP Licensing

**Viktor Braun and Cornelius Herstatt**

Routledge
Taylor & Francis Group
New York London

First published 2009
by Routledge
605 Third Avenue, New York, NY 10017

Simultaneously published in the UK
by Routledge
2 Park Square, Milton Park, Abingdon, Oxon OX14 4RN

*Routledge is an imprint of the Taylor & Francis Group, an informa business*



Typeset in Sabon by IBT Global.



*Library of Congress Cataloging in Publication Data*

Braun, Viktor R. G., 1982-
User-innovation : barriers to democratisation and ip licensing / Viktor R. G. Braun and Cornelius Herstatt.
p. cm.—(Routledge studies in innovation, organization and technology ; 13)
Includes bibliographical references and index.
1. Technological innovations. 2. New products. 3. Democratization. I. Herstatt, Cornelius. II. Title.
HD45.B657 2009
338'.064—dc22
2008043734

ISBN13: 978-0-415-77719-3 (hbk)
ISBN13: 978-0-203-87390-8 (ebk)

# Contents

# Diagrams

# Tables

# Frequently Used Abbreviations

| | |
|---|---|
| BDP | Bundesverband Deutscher Planzenzüchter [The German Association of Plant Breeders] |
| BMBF | Bundesministerium für Bildung und Forschung [The German Ministry for Education and Research] |
| cf. | compare/see also |
| CGIAR | Consultative Group of International Agricultural Research |
| CIMMYT | International Maize and Wheat Improvement Center |
| DMCA | Digital Millennium Copyright Act |
| e.g. | example given |
| EPO | European Patent Office |
| et al. | and others |
| EU | European Union |
| EULA | End-User License Agreement |
| FDA | US Food and Drug Administration |
| Fn. | Footnote |
| FAO | Food and Agriculture Organization of the United Nations |
| FTC | Federal Trade Commission |
| HYV | high-yielding variety |
| ICTs | information and communication technologies |

| | |
|---|---|
| IP | intellectual property |
| IPR | intellectual property right |
| MTA | material transfer agreement |
| NGO | non-governmental organization |
| OSS | open-source software |
| PVPs | plant variety protection |
| PVS | participatory varietal selection |
| P2P | peer-to-peer |
| R&D | research & development |
| RTLA | reach-through license agreement |
| Sec. | section |
| SMEs | small- and medium-sized enterprises |
| TC | Trusted Computing |
| TRIPs | Trade-Related Aspects of Intellectual Property Agreement |
| USPTO | United States Patent and Trademark Office |
| USC | United States Code |
| WIPO | World Intellectual Property Organization |

# Foreword

My good colleague, Professor Cornelius Herstatt, has long worked on the topic of user innovation. In his previous work, and in common with many other colleagues, Herstatt has participated in developing our understanding of successful innovation by users, and of methods by which the innovative capabilities of users can be enhanced. In this book, he and Dr. Viktor Braun explore a very different and very interesting complementary topic: They describe some of the *barriers* that can be put into place by companies seeking to diminish or restrain the ability of their customers to innovate by modifying the products they sell. Resistance to user innovation by some or many producers is certainly understandable: Traditional producer business models are often affected by user innovation, and producers that are negatively affected may seek to slow or thwart user innovation rather than adapt to it. Still, it has also been found that social welfare is enhanced by user innovation, relative to a situation in which only manufacturer innovation exists. Therefore, policymakers as well as scholars and business people might want to review the barriers to user innovation that the authors describe in the case studies in this book, with an eye to appropriate modifications of public policy.

One of the cases studied by the authors is of special interest in this regard. Monsanto is a major supplier of seed to farmers, and an important innovator in the genetics of seeds for food crops. Its products include genetically modified seeds and also complements to them. For example, Monsanto develops and markets seeds with traits that allow plants grown from them to survive the application of certain specialized herbicides—and also sells those specialized herbicides. Purchasing and using these seeds can be very valuable to farmers.

Farmers—seed users—have also traditionally been innovators with respect to the seeds they sow, and many very valuable new crop lines and variations have come from these user-innovators. Farmer innovation processes involve intentionally creating plant crosses, and also simply observing the traits of the plants in their fields and selecting desirable ones. After all, mutations occur and all plants in a field are not the same. If they see a plant with a trait that they think is desirable, they may save the seeds from that plant and sow them again next year, gradually build up their stock of the modified seeds if the trait breeds true, and so on.

Monsanto sells its seeds with proprietary modifications protected by patents and as a condition of sale prohibits farmers from attempting to modify "its" seeds further. Monsanto has been very successful in court challenges to its practices, and certainly is a very successful business at the time of this writing. But its practices also clearly create a major barrier to a very valuable form of user innovation in agriculture. Is the blocking of user innovation, as illustrated by the case just outlined, a socially desirable state of affairs? Personally, I think it is not. In my view, governments and policy makers should think about how to reduce the collateral damage to user innovation that can be created by governmental policymaking both within and beyond intellectual property law. This book is a very useful contribution to a very important and just-emerging debate.

Professor Eric von Hippel,
MIT Sloan School of Management
July 2008

# Acknowledgments

> "The world comes to us in an endless stream of puzzle pieces that we would like to think all fit together somehow, but in fact never do."
>
> Robert Pirsig in *Lila—An Inquiry into Morals*

While the world that we inhabit and the priceless combination of events that join to form our respective lives rarely if ever create a coherent comprehensive whole, it is relatively easy and fulfilling to express gratitude to those who helped shape this present work. First and foremost we would like to dedicate this book to our respective families. In the case of the junior author this refers to his beloved parents, who have helped spark and nourish his desire to continuously expand his horizons.

Secondly, we are greatly obliged to the passion, vision and inspiration of Eric von Hippel, who has, over the years hosted and mentored both authors at MIT's Sloan School of Management and whose work serves as the foundation for a more user-centered world.

While the impetus of modern individuality is to give yourself excessive credit, the true essence of all modern scholarly writing is cooperative. Very few if any ideas are truly original. As the long bibliography demonstrates, we have borrowed a tremendous amount from the works of (brighter) minds. Apart from the referenced authors the interaction with industry and academic colleagues, especially at Hamburg University of Technology, the Universidad Autónoma de Madrid and the Massachusetts Institute of Technology, has greatly contributed to the creation of this manuscript. We owe a special gratitude to David Probert, Professor Paloma Sánchez, Dr. Ruperto González, Dr. Hans U. Brauner, Christian Cancino, Frank Tietze, Raj Tiwari, Christoph Stockstrom and especially to Dr. Nizar Abdelkafi. Finally, John Brewton's insightful comments and Anneliis Svenson's passionate editing further enhanced the manuscript. Despite of all the feedback, all errors and oversights remain purely attributable to the authors.

A grant by the Deutsche Forschungsgemeinschaft (DFG) enabled the efficacious completion of this project. We are thankful for their support and hope that our work serves as the desired contribution to academic and intellectual progress.

It was also a pleasure to collaborate with the hardworking staff at Routledge.

Last but certainly not least we would like to thank all of those who deserve the special title "friend." Benefiting from your wisdom and warmth along the way has been a critical puzzle piece to provide a sense of purpose for our journey. Johann Custodis, Dominik Gisch, Florian Geistdörfer, Max Gliner, Abhijit Pandya, Alan Muhammed, Victoria Zamora, Lucia Woolley and Hildegard Fritz-Denneville have been especially important for the junior author over the course of the last years. Just because it seems impossible to coherently align the puzzle pieces of our lives does not give us an excuse for not trying.

Viktor Braun and Cornelius Herstatt
July 2008

# Introduction

The Chinese have an ancient curse that translates to the wishing of another party "to live in interesting times." It is ironic how a curse can as time goes by and attitudes change become a positive modus vivendi that stresses the value of life-long learning from experience. While the world that we currently inhabit can be described as unjust, discriminatory, wasteful, unstable or superficial, the term dull certainly amounts to a misnomer and merely reflects badly on its author. The 21st century is an extraordinarily intriguing period of constant flux in which humanity will change profoundly. The world is simultaneously moving closer together and breaking further apart. Post-modernity has become an anachronistic term of the past. Technology has always extended the breadth of what is possible and opened up new avenues for previously unthinkable activities and processes. What distinguishes recent changes is the increasing inclusion of the human himself. Over the last thousands of years of technological progress humans have learned to employ natural resources to feed and warm themselves, have created artificial habitats and applications to simplify their lives and enable social and professional exchanges. While hygienic instruments, medicines and nutritional advances have for a long time helped to prolong the life span of humans, recent developments in nanotechnology, genetics, biotechnology, robotics and implantology promise to change (advance or destroy depending on your view) the nature of the human being itself. Silicon breast enhancements, organ transplants or the addition of advanced prostheses are merely the beginning. The human greed for beauty, strength and intelligence will ensure that all parts of the human body, including foremost the brain, will eventually be improved upon technologically. We are on the verge of something very substantial: The evolution of the human species.

In this light we strongly believe that everyone should be concerned with technological progress. Despite of its societal importance and the increasing academic and industrial focus on the field, innovation, broadly invention plus diffusion, i.e. the implementation of technological advancement, is still insufficiently understood. The "black box," as Rosenberg (1982) succinctly refers to it, of how the new arises, enters and reshapes our world is an inherently complex and frequently idiosyncratic process. Innovation extends far beyond the scientific and technological realm. Instead it is dependent on and shaped by a wide variety of social, political, legal and economic factors. Our understanding of innovation therefore requires much more than just the individual technology and its diffusion, but instead a consideration of the overall environment which allowed it to prevail (or not).

Technological developments have not only vastly reformed our lives, from mobile phones to genetically enhanced frost- and disease-resistant crops, but also changed the innovation process itself. Breathtaking breakthrough innovations may originate from high-tech laboratories employing PhD-holding engineers, computer programmers or scientists utilizing cutting-edge instruments but their societal influence will always, at least partially, be determined by their societal acceptance. By its very nature innovation requires the acceptance of consumers. While to some extent end-users of products have always been involved in the development of new products and services, especially if they had the power and means to order solutions according to their specific wants or requirements, their impact has traditionally been underestimated. Novel information technologies and increasingly mobile communication platforms have already empowered users by granting them the ability to access and instantaneously exchange cutting edge know-how and programming tools. This development is the driving theme of this work: The possibility that users could eventually replace manufacturers as the preeminent source of innovation, as described in MIT Professor Eric von Hippel's inspiring *Democratizing Innovation.*

Under the most extreme form of such paradigm manufacturers would become redundant in the innovation process as all types of users would be capable of innovating for themselves. A somewhat less revolutionary scenario would be that mass-customization with its "one-of-a-kind products and services that match unique customer needs" becomes ubiquitous (Sioukas 1995). Even if neither of these possible evolvements seems likely in the near future, a growing amount of companies in numerous industries have recognized and accepted that customer involvement is a very valuable source for product and service innovation. Sophisticated market research techniques, such as the "lead-user method," are therefore increasingly being used by companies to understand the needs and wants of their (potential) customers and explore avenues to more effectively integrate these within their new product development work. Resource dependence theory, relationship marketing and new product success studies provide some theoretical and empirical background to support this assumption of customer involvement as a rising source of additional value.

But do these developments foreshadow an increasing amount of user and a declining extent of manufacturer innovation? What is the likelihood that such a major shift towards user or customer dominated innovation will materialize? Is the innovation landscape really becoming democratized? In other words, is user-innovation really becoming easier and therefore increasingly widespread? What are the limits of user involvement and the user's limitations to innovate? We assume that not all users can or will equally contribute to new product development. We further assume that their capabilities to innovate will vary greatly and that their behaviour will depend on their requisite incentives and expectations. What are the barriers that users face in their efforts to innovate? Are incumbent corporations embracing or combating innovative users? And are there any indications that it may even be becoming more difficult for users to modify existing products and create novel ones?

These critical questions remain largely unanswered. While discussions on the benefits of user-led innovation abound, the literature has largely ignored the difficulties that users encounter in the innovation process as well as the occasional negative impact of customer involvement from the perspective of a manufacturer. In fact, apart from the phenomenal success of open source innovation, our understanding of user managed innovation is still limited as to its applicability and benefit to other fields. Furthermore, it is uncertain whether users prove to be useful at all stages of the product/industry life cycle.

This book is intended to discuss these issues in more depth. If we wish to improve our understanding of where innovations originate from we have to strive to comprehend how users participate in and shape this process. In this sense we aim to shed some light on the potential democratisation of innovation. Through an analysis of the hurdles that users wanting to innovate face in this process we desire to contribute to a deeper comprehension of the future innovation landscape which in turn will substantially affect the world that our and forthcoming generations (in whatever shape or form) will inhabit.

The reaction and changing role of corporations in this road also merits attention. There is a lot to be said that the—in our opinion inevitable—technological evolution of the human species should be controlled not by a handful of large corporations, but that a diverse population of users, who are driven by different motivations than mere financial return to investment, should also have their fair share of influence.

The arguments in this book are based on an in-depth literature review, an empirical analysis, numerous interviews and various case-studies. This book is structured in the following manner. Part A introduces the user as an important source of innovation and outlines the democratization argument. Part B will provide both an explorative and empirical analysis of the main barriers to user-innovation. In Part C we will discuss the dynamics of user-innovation. In the final part of the book, D, we will introduce the importance of licensing in the innovation process, argue that current trends, especially technological ones, are not pointing towards a democratisation of innovation and suggest that licenses could serve as a tool to enable manufacturers and users to constructively collaborate.

# Part A
# The Democratization Potential

Our journey takes us into the competitive and complex world of business. Here capitalism, in some form or other, has been dominating in many countries since the early 16th century. Rather than merely being a term for the general economic system of our times, it also characterizes modern men since the Renaissance. The continuous will to improve oneself and to act rationally has led men and women of today to be referred to as "homo economicus," (relatively) reasonable and opportunistic individuals who often act according to their economic interests (Guckelsberger 2005). The strife to continuously enhance our standards of living is also one of the key distinguishing factors between modern men and the animal kingdom. While animals evolve through natural selection, our ability to preserve and convey knowledge to others along generations has enabled us to achieve something unique: Continuous technological progress built up over time.

At its most basic level capitalism refers to the sum of the economic decisions of each individual. Interestingly the vast majority of innovation studies largely ignore the individual in favor of focusing on the overall cumulative effect for the whole community. Regardless of whether we take an individual or cumulative approach however innovation affects every part of our lives from our professions to sports,[1] to academic science and even to religious movements.[2] Despite of the occasional outcry or protest, technology has ultimately prevailed in the vast majority of cases.[3] Few individuals today could imagine a social and professional existence without electricity, running water, mobile phones, the internet or automobiles. Both the modern individual and society as a whole have become dependent on technological progress. How else could we feed an ever-increasing global population that is projected to grow from the 6.5 billion people that inhabited our planet in 2006 to 9 billion by 2050?[4]

Technological advancements have not only changed our world in uncountable ways, they have also has shaped the playing field for future progress. Nobel-prize winning economist Robert Solow (1957) provided evidence for the assertion that the majority of the productivity increases in the US in the first half of the past century came from technological progress. Boskin and Lau (2000) confirmed that over half of the economic growth in the 20th century in Germany, Italy, the UK, France, the US and Japan came from technical improvements. Innovation has changed the business world and practically become a prerequisite for long-term commercial success. Ughanwa and Baker (1989) found that a decline in product innovation generally leads to a decline in market share. Trott (1998) explained that a main reason for the bankruptcy of the last British mass manufacturer of motor cars, Rover, was its inability to develop new innovative products. "Growing the business through innovation" was rated as the number one problem by US corporations since 2002 in the annual survey of the Industrial Research Institute (IRI 2007).

The term innovation however has no uniformly accepted definition. In fact the existing definitions are even slightly conflicting.[5] One of the first definitions of innovation, the one advocated by Joseph Schumpeter (1934) in *The Theory of Economic Development*, will serve as our starting point:

1. The introduction of a new good—that is one with which consumers are not yet familiar—or of a new quality of a good.
2. The introduction of a new method of production, which need by no means be founded upon a discovery scientifically new, and can also exist in a new way of handling a commodity commercially.
3. The opening of a new market—that is a market into which the particular branch of manufacture of the country in question has not previously entered, whether or not this market has existed before.
4. The conquest of a new source of supply of raw materials or half-manufactured goods, again irrespective of whether this source already exists or whether it has first to be created.
5. The carrying out of the new organization of any industry, like the creation of a monopoly position or the breaking up of a monopoly position.

This definition indicates the breadth of the term. It indicates that innovations can occur in a wide variety of areas. Our focus will be on product innovation and technological process innovations. Marketing, organizational or service innovations will not be considered. Schumpeter considered innovation to consist both of developing new technologies or finding new uses for existing ones and implementing these into existing or new products or processes. Without such implementation, we will not consider an invention, modification or improvement to be an innovation. Various commentators confirm this view. Roberts (1988; p. 581) formulates that

Innovation = Invention + Exploitation

Lundquist (2005) concurs and states that innovation consists out of

a) Invention, i.e. research & development (R&D)
b) Productization, i.e. design through manufacturing
c) Launch, i.e. market introduction

Veryzer (1998a) offers another related innovation model:

1) Concept generation and exploration
2) Technical development and design
3) Prototype construction
4) Commercialization

While there are many other models of the innovation process, those mentioned above clarify some important concepts. The first component of Lundquist's model, invention, is often used to refer to the results of organized scientific corporate research and development operations. Veryzer's model and Rosenberg's (1982) concept of "learning by using" indicate however that informal R&D processes, such as non-formal experiments by third-parties, can be sufficient. Subsequently the concept has to be turned into a product. A prototype has to be created and the product has to be prepared for production. Finally, the product has to be released on the market. Here the words launch or diffusion seem more appropriate than commercialization as not all innovations (e.g. open-source software programs) are commercially exploited. Whether or not launch suffices the offer of a new product at a local corner shop is an unanswered question. In the realm of agriculture, Schlebecker (1977; p. 647) tied innovation to commercial success: "if a new method or device has been adopted by about 10 percent of the possible users, then, possibly, a technological innovation has taken place." While such an approach seems somewhat arbitrary, it does highlight that innovation is a vague, and at times, even subjective term. For the purpose of this study, we define innovation broadly as the resolution of a problem such as by the improvement of a product plus its subsequent diffusion.

There are different types of innovation. The main differentiation is between radical and incremental innovation. Radical or discontinuous innovations refer to the "development or application of significantly new technologies or ideas into markets that are either nonexistent or require dramatic behaviour changes to existing markets" (McDermott and O'Connor 2002; p. 424). Incremental innovations are minor extensions or improvements to existing products or processes. While users have generally been considered to only generate incremental innovations[6] as their familiarity with existing products often limits them from thinking 'outside the box', Lettl et al. (2006) demonstrated that users can also play an essential role in the development of radical innovations. While this research is largely focused on incremental improvements it will also be applicable to radical innovations.

Identifying and studying where exactly innovations originate from has become an important area of research.[7] We will focus on the functional locus of innovation which categorizes innovating individuals and firms in terms of the functional relationship via which they derive benefit from a given product or process.[8]

Since the Industrial Revolution, the common belief has been that new product innovations result from large internal R&D operations. According to this traditional view, those organizations which partially or wholly create and commercially exploit hardware or software products[9] were exclusively responsible for the generation of innovations. Under the so-called manufacturer-active paradigm (MAP), manufacturers, producers or suppliers largely control the innovation process and are responsible for building, testing, producing and selling products.[10] This paradigm merely considers users to amount to purchasing consumers.[11] Ramírez (1999; p. 51) even explained that "in the industrial view customers [were considered to] destroy value created by producers."

While the logic to obtain monopoly rents seemed convincing until the late 1970s/early 1980s, in recent years the resulting conglomerates have struggled to survive. The modern economy is centred on speed and not size.[12] Corporate behemoths increasingly had to realize that the "golden age for internal R&D" where companies could collect the best equipment and the best people to create the best products by leveraging internal sources only and profiting from strong economies of scale to R&D is largely over (Chesbrough 2003). Modern information and communication technologies (ICTs), an increasingly mobile workforce[13] (e.g. in the EU across national borders) and the growth of the venture capital market, which has increasingly made finances available depending on a suitable business model rather than size,[14] have forced companies to rethink their management of industrial R&D. Companies are therefore recognizing that in an increasingly complex world it is impossible to do everything one self and that leveraging external research will be the key to future success.[15]

This changing nature of R&D and innovation, which has been termed the open-innovation paradigm,[16] has also increased the focus on the role of other parties in the innovation process. While universities, government institutions or suppliers frequently are involved in the development of new products, a relatively novel stream of research has persuasively argued that users have been and are increasingly becoming a critical source of innovation. Herstatt (1991) defined users as individuals (or a group of individuals) who use a product to achieve a certain end. The important aspect here is the functional relationship of the users with the product. End users are those entities that utilize a finished good. This ranges from the amateur racket swinging tennis player to the professional medical researcher employing a computer program to the seed-planting farmer who may be operating under the corporate veil.

As a counterpart to manufacturer-dominated innovation, Eric von Hippel (1976) coined the terms user-dominant innovation and customer-active paradigm (CAP) to refer to the scenario where the user is not just a passive consumer but rather an active creator. The innovation diffusion literature calls this process "re-invention": "The degree to which an innovation is changed or modified by a user in the process of its adoption and implementation" (Agarwala-Rogers 1978; p. 139). Re-invention or user-modifications refer to a) using the same technological innovation in a different manner than originally intended or b) enhancing the existing manner of solving the given problem.[17] User-innovation involves users developing a solution to a currently unmet need of theirs, creating a product on that basis and diffusing the results.[18] Re-invention will also be referred to as successful tinkering, experimentation or modification of a product.

Just like most black and white graphs consist of countless shades of the two colors, the functional locus of innovation frequently resides somewhere between the extremes of the MAP and the CAP. To provide some classification for this continuum Foxall and Johnston (1987) proposed four different degrees of user-innovations, ranging from simply solving an internal problem via a solution that is of no significance beyond the firm's borders ("user-initiated process innovation" and "passive user-initiated product innovation") to users securing IP rights and attempting to commercialize their developments ("vertically integrated user-initiated product innovation"). Voss (1985) also presented a model of the user-innovation continuum that is presented in Table 1.

It is sensible to separate active from passive user involvement and to distinguish contributions to the user-domain from those in the technological

*Table 1* Continuum of User-Innovation

| | *Dominant Party at Innovation Stage* | | | |
|---|---|---|---|---|
| *Category** | *Need Recognition* | *Idea Formulation* | *Development* | *Commercialization and Diffusion* |
| 1. User developed | User | User | User | User |
| 2. User developed, transferred to supplier | User | User | User | Supplier |
| 3. User innovation, supplier developed | User | User | Supplier | Supplier |
| 4. User initiated, supplier innovation | User | Supplier | Supplier | Supplier |
| 5. Supplier innovation | Supplier | Supplier | Supplier | Supplier |

Source: Voss (1985)
*While the first three levels are "user-active", the next two are "supplier active".

domain.[19] While an active contribution involves the development of own-solutions to a problem, passive contributions are made by providing innovation-related feedback. Contributions to the user-domain require only use-related know-how, while requirements in the technological domain require technological competences as well. As an extension of this view, independent user-innovation requires an active development contribution to the technological domain in addition to the ability to diffuse the development in question.

As a caveat, economic actors often hold different roles not only in relation to different counterparts but even in relation to the same counterpart.[20] While one corporation may use the other's products in one area, in another related field it may be the other way around. The decision whether an innovation is generated by a user or a manufacturer may thus frequently be a difficult one involving subjective judgements. It further indicates that users can contribute to all stages of the innovation process,[21] and that users and manufacturers can collaborate and interact with each other in many different ways. In some cases a user may initiate the developmental process, while in others manufacturers may take on this role and search for users. This can enable manufacturers to acquire novel technological competencies, remain at the cutting-edge of technological trends and "extend their innovation and technology-related networks" (Lettl et al. 2006; pp. 263–264).

Of the numerous ways of integrating users into new product development efforts,[22] the lead-user method, depicted in Diagram 1, is especially

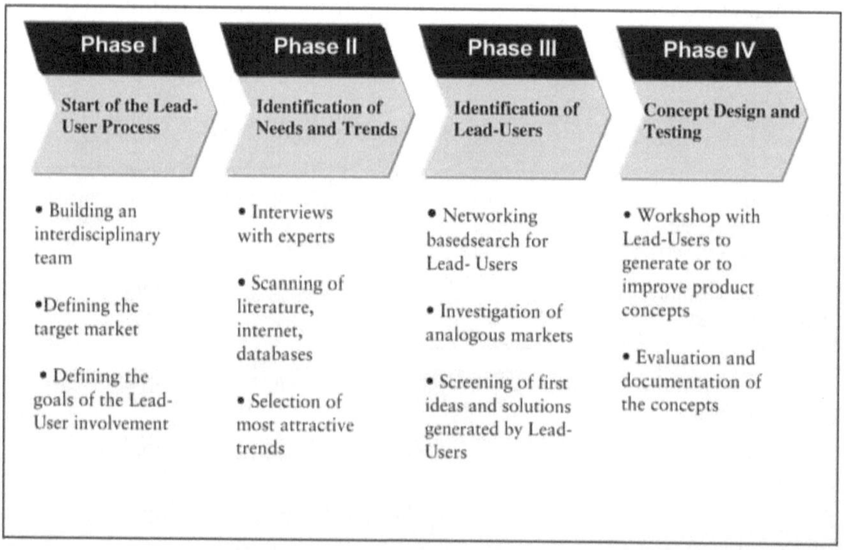


*Diagram 1* The process of the lead-user method.
Source: With permission from Herstatt 1991/ Lüthje and Herstatt 2004

noteworthy. Most of the customers typically involved in classical user need-assessment surveys appear to be unable to free themselves from the current product generation, the so-called "functional fixedness" (von Hippel 1988), and are therefore of little value in predicting future market needs. The lead-user method can solve this dilemma by allowing manufacturers to identify and cooperate with leading-edge users, who often have radically innovative ideas which they have occasionally even pursued to a prototype stage and beyond. The interested reader can refer to the case study in appendix 1 for a detailed operation of the method. Lead-users are "ahead of the majority of users in their populations with respect to an important market trend and they expect to gain relatively high benefits from a solution to the needs they have encountered there" (von Hippel 2005; p. 4). These two lead-user characteristics have been supported by theoretical work and empirical findings.[23]

It is critical to point out that the lead-user status is dynamic and changes over time.[24] The past innovative contributions of a user are not necessarily an indication of a future ability to innovate. Our task is to look at the dynamics of the innovation process in respect of users. Numerous studies have demonstrated that the traditional view of users as "passive acceptors of an innovation, rather than active modifiers" (Rogers 2003; p. 180) is flawed. The modification of an innovation through usage has been accepted to be "a natural part of its implementation" (Rice and Rogers 1980; p. 504). Franke and Shah (2003) and Herstatt and von Hippel (1992) demonstrated that in the areas of extreme sporting equipment and pipe hangers respectively well over 35% of users had developed innovative adaptations and applications.

This is as technology always has to be considered from the perspective of its social use:[25] Users "can embed new meanings into the technology during the course of that use" (Kline and Pinch 1994). The unexpected success of the Short Message Service (SMS) and the use of early automobiles by farmers to power agricultural machinery and thereby to transform farm life in the US illustrate this.[26]

Studies have found that in both consumer and in industrial goods markets users have frequently been a significant source of innovation since the industrial age.[27] Schiffer (2003) for instance describes how electrotherapists during the 1700s amended the devices used in physical instruments as these were too clumsy to suit the needs of their patients and thereby developed hundreds of different electrical appliances. Adam Smith (1776) even stated that "a great part of the machines made use of in those manufacturers in which labour is most subdivided, were originally the invention of the common workmen, who, being each of them employed in some very simple operation, naturally turned their thoughts towards finding out easier and readier methods of performing it" (cited in von Hippel 2005b; p. 65).

Farmers, historically probably the largest group of user-innovators, have for centuries been responsible for yield improvements in agriculture.[28]

Athletes have created an abundance of different sporting activities using countless different devices that they largely developed themselves. In recent years the sports of mountain biking, kitesurfing or rodeo kayaking were invented by users.[29] In the medical field, surgeons and doctors have been very active in the creation of numerous novel medical devices, treatments and in finding alternative uses for existing prescription drugs.[30] A study in the 1960s found that almost all of the most important oil refining innovations were developed by user firms.[31] Knight (1963) documented that in the 1940s users played a considerable role in the gradual evolvement of the computer hardware industry. TipEx was invented at the end of the 1950s by a secretary. The invention was subsequently taken over by 3M and implemented on an industrial scale. Freeman (1968) indicated that the most widely licensed chemical production processes were developed by user firms. The sport's drink Gatorade was developed by the trainer of the college football team of the University of Florida. Innovative users have also been encountered in areas as disparate as transport, toys, electronic circuit boards, libraries, scientific instruments or construction equipment.[32]

The main reason why in certain situations users are the more suitable innovators, is that they adopt a technology with their idiosyncratic needs which may give rise to a new application of the technology in question.[33] The user-innovation literature has pointed out that users innovate as homogenous or standard products are not meeting their heterogeneous needs.[34] The problem for manufacturers is that user needs, due to their intangibility, are "sticky" and therefore inherently difficult and expensive to transfer (Ogawa 1998, von Hippel 1994).

From the mid 1970s onwards, since Professor Eric von Hippel of the Massachusetts Institute of Technology became the strongest proponent of user-innovation, the interest in the phenomenon has systematically increased. Denmark for instance has become a hub for both user-innovation research and practice. The UK has amended its innovation policy to implicitly take the importance of embracing and encouraging user-innovations into consideration. Supporting a widening of the players who participate in the innovation process will be one of the main challenges of national innovation policies in the forthcoming decades. If users were able to modify most of the products they use or create completely novel products themselves that are specifically tailored to their needs, the risks of technological progress could be substantially lowered. The digital divide and the societal risks that a further rise of the already huge difference between the affluent and the destitute pose are obvious.[35] The substitution of human labor by machines and robots in countless realms, from the production of products to increasingly the field of services (e.g. automated telephone hotlines) and one day even in the creative realms and in R&D constitutes another considerable risk. Ray Kurzweil (1999) and various other artificial intelligence experts[36] even believe in the exponential growth of innovation, in which machine intelligence will surpass human intelligence to make machines the innovators of the future. They contend that such development has the potential to enable a small elite, those in charge of the innovating machines, to control the future of technological progress.

Both of these dangers could be mitigated if a growing number of users could control and improve the technologies they employ. If however technological progress and the resulting technological platforms are shaped and controlled by a few players the potential of technocratic abuse could threaten the fabrics of our society. The large scale crackdown on freedom of speech in China, as recently demonstrated by their substantial limitation of online videos,[37] serves as a brief illustration of this threat.

Is it becoming easier for users to participate in the innovation process? Global access to an unprecedented amount of information enabled by the explosion in information- and communication-technologies, user-networking through virtual communities,[38] and the wider access to information and simple programming languages,[39] have shifted power from "large institutions to individuals" (Shapiro 2000; p. 10).[40] The most influential aspect of the internet for users was the ability to instantaneously share digital content and technologies. This reshapes the innovation process as it enables the "massive diffusion" of information goods such as software (Dahlander 2006; p. 20), and thereby dramatically facilitates the independent exploitation and distribution of users' own developments. The open-source movement with its successful developments of Apache, Sendmail or Firefox illustrates this potential. The kernel to the most prominent open-source program, Linux, was created by over a thousand developers world-wide, with further tens of thousands shaping the overall Linux operating system.[41]

These developments have prompted some commentators to voice very optimistic outlooks. Neff and Stark (2004; p. 175) for instance explain their vision of a "permanently beta" business environment, where technologies are constantly in a state of flux and subject to "a process of negotiation [with] users." Harvard Law School Professor Jonathan Zittrain (2005; pp. 18–19) comments that the "widespread dissemination of computers with multipurpose operating systems spawned a Galapagos-like ecosystem of software designers," indicating the potential exponential growth of user-innovators. Prahalad and Ramaswamy (2000) argue that we are moving towards a world that will be characterized by user and manufacturer "co-creation" where users, at least partly, want to shape the experiences that products offer them. Ramírez (1999; p. 59) even suggested that "systems must be installed to track what happens to the customers' value creating." Eric von Hippel (2005a; p. 1) summarizes these views by persuasively arguing that that innovation is being "democratized" and "that users of products and services—both firms and individual consumers—are increasingly able to innovate for themselves," independent of manufacturers.

This assertion however lacks empirical foundations. It is extraordinarily difficult to answer whether the majority of innovations result from the work of users or from manufacturers. Similarly, to determine whether or not user-innovation as a whole is becoming easier will depend on the scrutiny of a wide variety of factors, such as the type of technology involved, the market structure, geographical issues, legal considerations, etc. Industry differences are likely to be considerable as a comparison between the open-source and the pharmaceutical industries illustrates: How many user-developed drugs have there been in recent years?[42] It is not our premise to find a conclusive answer to this question. Our goal is merely to further the debate in this area by looking at the other side of user-innovation.

Namely we will scrutinize the factors adversely affecting user's ability to tinker and innovate and therefore to critically assess the democratization argument. While the developments in Information and Communication Technologies (ICTs) can increase the potential of users to innovate in certain areas, it should not be forgotten that "many barriers . . . must be overcome in order to bring any product or process to commercial reality" (Shaw 1985; p. 48). We will argue that some recent developments, such as the global strengthening of intellectual property rights or technological developments in genetics or nanotechnology, have the potential to burden rather than to facilitate user-innovation. In this process we will affirmatively answer the question posed by Lüthje and Herstatt (2004) whether manufacturers can actively change the users' cost and benefit expectations with respect to innovations.

Our investigation will be relatively one-sided as the factors that could increase user-innovation will not generally be discussed as these have been pointed out elsewhere.[43] An in-depth discussion, in all its complexity, of all the relevant factors and their interactions is beyond the scope of this work. While it is accepted that only those users innovate who have a high need to do so and who operate at the leading-edge of their field,[44] meeting these criteria is however no guarantee that a user can innovate. Even leading-edge users in a field may not be able to innovate. Briefly consider the example of professional racing car drivers. Although they are the best users of a complex technical gadget, few, if any, would in themselves, without the help of a highly-qualified team of engineers, be capable of improving the performance of their vehicles. On the other hand, even users who neither have a high need to innovate nor are at the leading-edge of their field have been found to innovate. Mollick (2004) for instance describes a particular type of underground users who innovate without authority within closed proprietary technical systems and which have created product extensions of computer games and software upgrades to the TiVo personal video recorder.

# Part B

# Barriers to Democratization

# B.1 Exploring the Barriers to User-Innovation

## PREPARING FOR THE EXPLORATION

Technology can transform our society in dramatic ways, as the development of electricity, railways, the steam-engine or the printing-press have illustrated. What Schumpeter (1968; pp. 25–26) first termed "creative destruction" involves the giving up of existing social and economic structures to embrace the new. Overcoming significant resistance to change may thus often be a difficult political and social process. In the long run the success of a technology will frequently depend not necessarily on its merit, but rather on its social use and acceptance. The dynamics of numerous political, societal, legal and economic factors therefore have to be analyzed to adequately assess the long-term prospects of a forthcoming technology.[1]

The prevalence of user-innovation is, akin to technological success, not inherently inevitable. Berger (1975) and Boyden (1976) in their studies of engineering polymers and chemical additives for plastics for example found that 100% of innovations had been generated by manufacturers. Veryzer (1998b; pp. 143 and 148) found a low or "rather limited" role of users in regard to radical innovations. This was especially during the initial stages of the development processes as "secrecy concerns and the proprietary nature of these products restricted the amount of information that could be provided to customers" and as "it was not always easy for customers to comprehend or appreciate the radically new products or the[ir] implications." The shift from user-dominated to manufacturer-driven innovation that Knight (1963) and Baldwin et al. (2006) observed in the computer hardware and rodeo kayak industries respectively also indicates that user-innovation has its limitations and that its prevalence depends on numerous factors.

A study by Freel (2003) in the UK showed that while positive links for product innovation were found form co-operating with customers, internal R&D sources were the most important determinant of successful product innovation.[2] Nelson's (1993) edited work on *National Innovation Systems* confirms that in most industries and nations the majority of innovations are made internally. Spann et al. (1991) also found that in-house skills and efforts in developing and improving technology were the most important source of technical know-how. A recent study of 51 electronics firms in the UK and Ireland found that while customers were by far the most frequently contacted external organization much more than universities, suppliers or competitors, overall co-operation with users were found to have a marginal influence on the success of the innovation process.[3] A prior study of 1700 UK plants, 1300 German plants and 400 Irish companies similarly found that "networking [with third parties such as users] appears to have no positive role to play" (Love and Roper 2001; p. 657). Involving users in the R&D process therefore seems to be a question of strategy and may or may not be appropriate depending on the specific circumstances at hand.

While the user-innovation literature has so far argued that if user needs are heterogeneous then users will be more suited to innovate than manufacturers,[4] this argumentation seems to neglect their actual ability to innovate. Another explanation for the differences in user-innovation is von Hippel's (1981; p. 95) contention that the "functional locus of innovation can be effectively modelled in terms of appropriability of innovation benefit." In other words, the party which is most likely to appropriate the benefits of innovation will invest the time and resources. While interesting, the argument is based on numerous assumptions which may not be fulfilled in the real world. One of the necessary conditions that has to be met is that "innovators should not be able to effectively capture benefit from the licensing or sale of non-embodied knowledge regarding their innovation to others" (von Hippel 1981; p. 98). In the years since this study[5] however a sizable global "market for technology" has developed which according to different estimates varies somewhere between $35 billion to $150 billion per annum.[6] While there is clearly a lack of comprehensive empirical data over time, Arora et al. (2002; p. 2) conclude that "all the available evidence suggests that the trade in technologies is more common than it was in the past." While patents may not be the most effective tool to appropriate innovation benefits, they have been found to stimulate R&D across almost all manufacturing industries and both small and large parties have benefited substantially from licensing their intellectual property.[7]

At this stage it should be emphasized that any innovation, at a minimum, requires the "motivation to engage in the costly problem solving activity" (Lotz 1992; p. 8) and the technical ability to do so. The creation of new works also "requires access to and use of pre-existing data and works" [own emphasis added] (Dutfield and Suthersanen 2004; p. 379), know-how, tools, and other resources, such as finances. Overall, innovation involves both the right, a legal question, and the ability, an economic, social and technological question.

For users the uninhibited tinkering or experimentation is especially important.[8] Shulman (1999; p. 160) persuasively explains that the main reason why Edison was so successful was precisely due to his continuous tinkering and experimentation: By "trying out every design he could conceive of" he was able to develop such a vast amount of ground-breaking technology. Tim

Berners-Lee (2000) developed both the Hypertext Transfer Protocol (HTTP) and the Hypertext Markup Language (HTML), two important foundations of the world-wide web, by tinkering around with hypertext.

Productization and diffusion of non-information goods will also rarely be an easy endeavor. Although the majority of user-modifications and developments will be efforts to tailor the product to their own needs, where there is no necessity or desire for productization and diffusion, in some cases improvements may be relevant for other users. In such situations turning the improvement into a diffusible product, producing and launching it is likely to be problematic for most users.

The argument that will be made in this section is that there are various factors that have the potential to burden tinkering and to increasingly restrain both the access to information and the requisite tools. How could serendipity strike without the possibility to experiment? How can you discover a new use for a product if you can only utilize the device for the exact purpose it was made? As a brief anecdotal example of the importance of this research Princeton University Professor Edward Felten created a website to discuss the law and regulation surrounding "your freedom to understand, discuss, repair, and modify the technological devices you own."[9]

While the weakness of the manufacturer-active paradigm, i.e. the flawed assumption that users can only pass need-related but not solution-related inputs to manufacturers, has already been pointed out by various parties,[10] the CAP and the sweeping argument that user-innovation is being democratized have not yet been subjected to closer consideration. While user-innovation advocates continuously point out that the vast majority of manufacturer-developed products fail in the market place,[11] we were unable to identify a study of how and why many user developments fail. In light of the flourishing publications in this realm it does seem odd that there hardly exist any academic or practitioner writings in respect of the limitations of this paradigm.

In order to attempt to fill this vacuum and analyse the barriers to innovation a suitable framework is required. In *Code and other Laws of Cyberspace* Lawrence Lessig (1999) proposed a four-prong framework that allowed him to critically assess the developments of the regulation of cyberspace:

- law
- market
- architecture
- norms

As it has already been suggested that this broad framework is well suited to analyse the regulation of technology in general,[12] we will apply it in a slightly modified manner to evaluate and describe barriers to user-innovation. We will therefore assess the various legal, economic, technological (i.e. architectural) and social (i.e. norms) barriers that can prevent users from tinkering.

This framework is based on the realization that innovation is dependent on and shaped by a wide variety of social, political, legal and economic factors. Changes in technology will often eventually lead to legal adaptations. The legal reaction to the duplication technologies of the digital age, namely the vast increases in the protection for intellectual property, illustrates this. The law however should also be seen as a political process as it is shaped not only by the vague best interest of society, but also by the influence of powerful political pressure groups, the economic power of corporations and social norms. The law therefore not only shapes the other three modes, but it is also shaped by them. The same goes for the other three principal factors. Any major change will sooner or later affect the other modes, leading to a continuous cycle of mutual interaction. Although the barriers may be powerful individually, only their combined effect matters in practice.

In order to understand innovation we thus need to look at much more than just the individual technology and its diffusion, but instead at the overall environment which allowed it to prevail (or not).[13] In this quest to evaluate the combined effects of these forces we have to acknowledge that as there is little or no reliable statistical information available or feasible to obtain. Many of our assertions are therefore based on anecdotal evidence.

This part will commence with a short section outlining the corporate incentives to control, limit and prevent user-innovation. Subsequently, each individual barrier—legal, economic, technological and social—will be discussed in depth. The combined effect of the barriers will be seen in the empirical study of computerized dentistry in Part B.2.

## CORPORATE INCENTIVES TO PREVENT USER-INNOVATION

Before scrutinizing the barriers to user-innovation, we need to consider the motivations corporations have in controlling user-modification. The simple answer is that a democratization of user-innovation is not in the interest of incumbent corporations as this would endanger one of their core competences: the ability to innovate. Innovation, as the frequent use of the word in companies' Annual Reports indicates, is a key source of the competitive advantage of many corporations. For the last 6 years 'growing the business through innovation' was rated as the number one problem in the US Industrial Research Institute's annual survey.[14] While collaborating with users may in many situations be a promising strategy, it also has its risks and limitations.[15] It is therefore crucial to note that it is a strategic decision whether or not it makes sense for a corporation to collaborate with users. As different CEOs at 3M had different strategic foci the lead-user method was first successfully employed, then dropped only to be reemployed again.[16] A Scandinavian software firm seized using the lead-user method despite its successful applications due to a mixture of organisational inertia and the large amount of resources required to entrench the method in its organizational routines.[17] Gassmann (2006; p. 223) therefore explains that "open innovation is not an imperative for every company and every innovator." Companies should therefore strategically evaluate which problems are best solved via an open- and which via a closed-innovation approach.

While some barriers can directly be erected by manufacturers (such as through market power, contracts, IP or technology), others (such as the legal and social environment) can be indirectly shaped, by lobbying for example. The two main underlying reasons why erecting hurdles to (user)-innovation in these areas can make strategic sense are the economic one of increasing revenues and the strategic one of protecting business secrets. Granstrand (1999; p. 399) therefore explains that new technologies "enable investors and producers to build electronic locks and fences around their properties in order to generate and control their rent streams."

Basic economic theory promulgates that by charging different customers different prices, for instance depending on frequency of usage, companies can maximize their profits.[18] Aftermarket strategies of under-pricing hardware, e.g. razors and printers, and over-pricing, aftermarket components, such as razor blades or printer cartridges, exist in a wide variety of markets.[19] Technological barriers are necessary in order to exclude non-authorized reuse of their product and implement a market segmentation strategy. Microsoft for example attempts to limit the use of its software to one computer.[20] Printer manufacturers such as Hewlett-Packard or Lexmark[21] try to restrict users to only utilize printer cartridges once in order to appropriate as much as possible from the over $10 billion ink cartridge aftermarket.[22] If users could modify their technology themselves, such as by re-filling their ink cartridges, the business model of under-pricing hardware to benefit from the, often very lucrative, sale of after-market applications or replacement components sales would no longer work.

Douthwaite (2002) generalizes that corporations in most situations, due to their profit-maximization incentives, are more likely to introduce technologically complex products that are difficult to modify rather than simple solutions that can easily be shaped by users themselves. Additionally, user-modification could enable users to tailor products to their needs, which manufacturers may not desire as they may want to benefit from the provision of such service.

The second principal reason refers to the protection of companies' intellectual property. As the ability to innovate that comes from the generation and effective employment of new know-how is often a core competence of many corporations, these will be keen to protect the potential advantages that derive from it. Companies therefore often try to increase the difficulty of reverse engineering[23] as this allows rivals to appropriate such know-how.

Car manufacturers for example often make car parts interdependent and in one closed, sealed unit so that when one part is broken, the whole unit has to be replaced. To illustrate, the bonnet of the Audi A2 is intended to be opened only by authorized Audi dealers. This makes it exceptionally difficult for non-authorized parties, such as independent car repair stations or users, to repair such units.[24] Such strategy of limiting modification to authorized dealerships can allow car companies to both increase their rents and protect their know-how. Developing technologies in such a manner that they cannot be modified or copied is a strategy that is becoming increasingly important, especially in developing countries. As product pirates are becoming more and more sophisticated companies should employ all available tools to protect their know-how and competitive advantage.[25] Apart from preventing software piracy[26] the creation of tamper-resistant products is also critical in order to provide services such as banking transactions, on-line gaming, electronic voting and digital content distribution without the risk of intellectual property theft.[27]

Another important reason why companies intend to create non-modifiable products is the desire to evade legal liability, and the possible adverse effects on reputation, for damages that may result due to products being modified in an imperfect manner or used for non-authorized purposes.[28] Maintaining control over the quality of their product is a crucial consideration why user-innovation is often not desired by manufacturers.[29] To illustrate, take the example of a flatbed grain dryer developed in Vietnam, which was subsequently modified and reconstructed by local farmers. As these made mistakes in reducing the airflows of the blower, the excessively hot dryers caused the grains to crack, considerably lowering their sale value.[30] Another reason not to allow users to modify the product may be out of the fear that this would send an adverse message to the market of the product's quality.[31]

In the realm of products employing software, another goal is to prevent viruses or malicious hackers from modifying software programs, as this can possibly result in great loss to legitimate users and third parties. An extreme example of what can happen if tinkering is allowed is provided by the Islamic terrorists that recently attempted to blow up various airplanes using their iPods to ignite home-made bombs.[32]

Structural factors as well as the egocentricity of some manufacturers and diffusion agencies which are frequently of the opinion "that they know best as to the form of the innovation that users should adopt" (Rogers 2003; p. 184) also play an important role. Interacting too closely with users can furthermore "limit the firm's room for strategic action" (Dahlander 2006; p. 41). Once a firm allows users to freely tinker with its products a strategy reversal is likely to lead to very strong adverse reactions.[33]

Douthwaite (2002; p. 57) mentions a further fascinating reason why corporations may be keen to erect certain hurdles to innovation: "erecting a barrier or two favors the self-selection of the right sort of people." In other words, by imposing certain difficulties on users' ability to innovate, corporations can identify lead-users with whom collaboration may be sensible.

To summarize, whether to allow, stimulate, burden or prohibit user-innovation in a specific realm will always be a strategic consideration for the company in question. How to achieve the general goal of long-term commercial success will vary from case to case. Riggs and von Hippel's (1994; p.460) finding that in the scientific instruments industry there is "very significant tendency for users to develop innovations that have high scientific importance, and for manufacturers to develop innovations that have high commercial importance," seems to offer support for this view. In areas where the commercial benefits are expected to be low, manufacturers are likely to allow or even stimulate users in their innovative activities. In areas, where the expectations of commercial benefit are very high, manufacturers will be much more inclined to control user-innovation. Whether user-innovation will be embraced or limited will also change over time. Take the example of the Ford Motor Company. In the beginning of the 20$^{th}$ century, it not only permitted users to tinker with their Model T automobiles, but substantially benefited from it and used the slogan "The Universal Car" quite literally to its advantage: Model T automobiles were being used by tinkerers, mainly farmers, to run washing machines or to plow fields. Such embracing of user-innovation soon seized when Ford changed its strategy to not only sell Model T vehicles but also trucks and tractors. As user-tinkerers were no longer needed to increase Ford's popularity and instead threatened the sales of its new models, the company started to discourage tinkering: It gave out warnings that tinkering would damage the car's engine, drafted strong warranty provisions and exerted pressure on dealers not to sell truck or tractor kits to farmers.[34]

## LEGAL BARRIERS TO USER-INNOVATION

Law is frequently considered to be the most important means to regulate society. The law and its enforcement, especially in terms of intellectual property rights (IPRs), have a high potential to adversely affect user-innovation. Throughout the last century, "as information production has become more central to our economy, we have seen a Cambrian Explosion of exclusive private rights in information," as Harvard Law Professor Yochai Benkler (2002; p. 81) explains.[35] While in the US the Copyright Term Extension Act 1998 (CTEA) extended protection for copyrighted works by a further 20 years, patent rights were broadened globally to cover genetic material, business methods and even software.[36] Additionally, strong database rights were introduced[37] and powerful legal tools were erected to prevent the circumvention of technical measures that protect copyrighted works.[38] Finally, the Trade-Related Aspects of Intellectual Property (TRIPs) agreement internationalized intellectual property protection through the world and substantially enhanced the enforcement of these rights. TRIPs effectively forced all countries wishing to benefit from the WTO's advantages to implement IPR protection systems that are monitored and partly enforced through the WTO.

The global strengthening of IPRs is creating substantial obstacles for user-innovators. While in many case IPRs will not prevent users from generating improvements, especially for personal use, their widespread, can impose considerable limitations at the diffusion/commercialisation stage. We argue that the continuous trend towards ownership of the tools needed for tinkering, such as information (copyright, databases) or inventions (patents) endangers free user-innovation.

We will focus on the effects of patents and copyrights on user-innovation and subsequently briefly describe how the law may curtail the activities of hackers and database users. We will end the section with a discussion of why IPRs are likely to remain strong around the world over the foreseeable future.

## Patent Law

Due to various recent global developments in strengthening patent law around the world,[39] this age has been termed the "pro-patent era," which is characterized by the strong enforcement of patents and further upward international harmonisation efforts. The Trilateral Co-operation between the US Patent and Trademark Office (USPTO), the European Patent Office (EPO) and the Japanese Patent Office (JPO) provide statistics for the global demand for patents as measured by the total amount of global

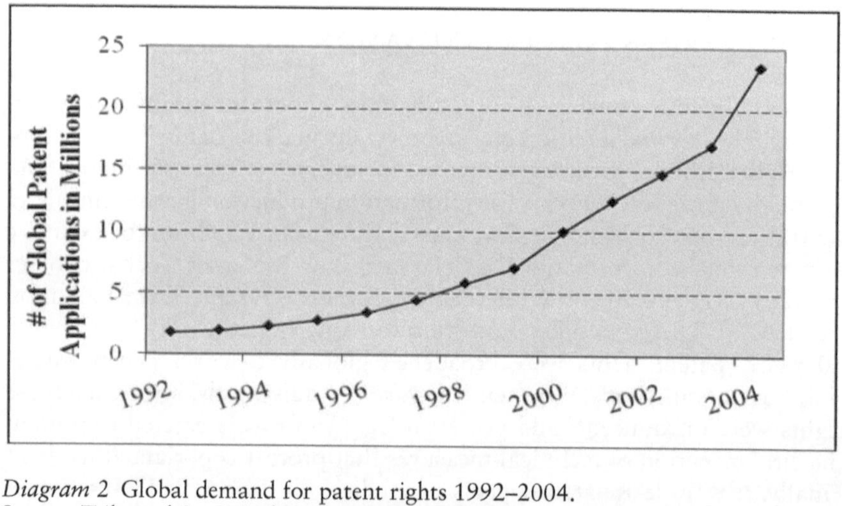


*Diagram 2* Global demand for patent rights 1992–2004.
Source: Trilateral Statistical Report 2005; 2000; 1996

patent applications. The huge increase since the early 1990s, visible in Diagram 2, indicates that patents have never been more fashionable.

Diagram 3 further reinforces this by showing the increase in the number of US patents that have been granted since 1900. The increase has been especially sharp over the last 20 years. While in 1984 67,200 patents were issued, by 2002 over a 100,000 more—around 167,331—were granted (USPTO 2006b). This amounts to an increase of around 150%. A study by the National Research Council (2004) further found that in the US the patents per R&D dollar had almost doubled between 1985 and 1997, increasing from 0.18 to 0.34 patents per million dollars.

We will now describe why Baldwin et al. (2006)'s statement that users "have the right but not the obligation to solve problems in a new way" may be, strictly speaking, incorrect. If users' innovative activities violate the patent rights of third parties, they may be obliged to seize their actions. Patents are granted on new technological developments that, subject to marginal national differences, meet the following criteria: novelty, non-obviousness, originality and susceptibility of an industrial application. Patents are national rights that allow their holders to exclude anyone else from making, using and selling a specific invention for a limited term, typically 20 years from the application filing date, in a specific territory. During such term a patent holder can largely control the use of the patented technology by for instance suppressing it, incorporating it into internal processes/products, licensing it to third parties or completely selling the rights to it. Clearly the use of a patent is always limited by the patent rights of others and other applicable laws, which cannot be violated in the patent's exploitation. While patents protect an owner against both independent invention/development of the involved technology and reverse

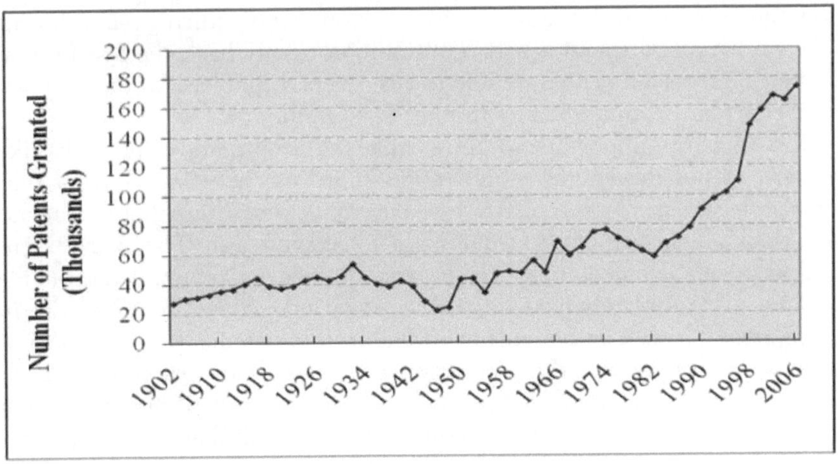


*Diagram 3* Global demand for patent rights 1902–2006.
Source: Based on data from USPTO 2008, 2006b

engineering by third parties, prior commercial use is a generally recognized excuse to patent infringement.[40]

There are two principal ways in which patents can burden user-innovation. The first is via limiting the area in which users can develop creative solutions, which we will discuss here. The second is through attached licenses, which we will discuss in the section on contractual barriers below. This refers to the crucial difference between a license and a sale. While with an ordinary sale a user can tinker as much as he likes with a product, including taking it apart, this may not be possible in case of a licensed use. While in regard to the former the user is only limited as to the creation and commercialisation of subsequent products that violate the patent, the license can prohibit such tinkering in the first place.[41] Software for instance is generally sold in the form of licenses that forbid modification and reverse engineering.

Patents, as illustrated by Carrier's (2003) tripartite innovation model, consisting out of the creation of the product, the recovery of the investment incurred in creating the product, and the circumvention of patent bottlenecks that block the path of innovation, generally operate at the diffusion/commercialization stage. While there is no legal obligation on innovators to undertake a prior patent search, it is prudent to do so from a business perspective.[42] Patents have been compared to landmines: "The easy way of finding them is not the good way of finding them" (Libin and Gray 2003). Ignoring other parties' patents can have immense consequences, such as the shutting down of a company's business.[43] Knowingly infringing a patent can even result in being liable for treble damages, plus attorneys' fees, as well as potential individual liability for a company's officers and directors.[44]

The most notorious case on point is Polaroid's successful suit against Kodak.[45] The later was ordered to pay $909 million for patent infringement and was barred from competing in the instant film camera business. Kodak consequently had to shut down its $1.5 billion manufacturing plant, lay off 700 workers, and spend nearly $500 million to buy back the $16 million instant cameras it had already sold to customers. Legal fees for the 14-year-long court battle amounted to another $100 million.[46]

It is therefore important to know what the patent landscape looks like as early as possible in any modification/improvement effort to avoid subsequent legal surprises. To clarify, we have to acknowledge that patents do not block all subsequent innovations in a specific realm. The exact extent and scope of a patent is fuzzy, i.e. its exact boundaries are not determinable ex-ante, and ultimately have to be decided by a court. While in practice in some areas[47] it may be possible to invent around patents, this requires not only prior awareness and understanding of the relevant patent rights, but also an additional input of resources.[48] Since the early 1980s navigating through the patent thicket has become a difficult task that requires professional dedication and legal support. Eighty-nine percent of the respondents to an OECD survey for example stated that the business risks of not patenting an invention had increased over the last 10 years.[49]

The example of the US is especially enlightening. Ever since the establishment of the Court of Appeals for the Federal Circuit (CAFC), which hears all patent appeals and changed the previous sceptical judicial view of patents to one that promoted broad, exclusive patent rights, and the federal antitrust laws in relation to patents were relaxed,[50] patents have become a burden for user-innovators. Not only did the total amount of patents in the US triple from 1982 to 2002, when nearly 170,000 were issued,[51] but the way in which these were enforced also changed. While the courts previously rejected the majority of claims against alleged infringers,[52] the policy changes led to a sharp increase in plaintiff success rates.[53] Jaffe and Lerner (2004; p. 2) thus explain that

> the new court of appeals has interpreted patent law to make it easier to get patents, easier to enforce patents against others, easier to get large financial awards from such enforcements, and harder for those accused of infringing patents to challenge the patent's validity.

Companies therefore increasingly started to strategically patent,[54] building coherent groups of patents around their inventions, so-called patent fences, to prevent competitors from being able to develop rival products.[55] Reitzig (2004; p. 473) in an in-depth study of European patents found that on average inventions are protected by a "coherent group of around five patents." Simultaneously, companies such as inter alia Intel, AT&T, IBM and Motorola adopted more aggressive patent enforcement policies,[56] resulting in a substantial increase in US patent disputes.[57] McGavock and Lasinski (1998), in a survey of the infringement litigation policy of 186 Licensing Executive Society members, found that only 3.4% of respondents ignore minor infringers (that is if they find out about such infringements). 20.7% of large companies, those with more than $500 million in revenues, responded that they would "litigate at all costs."

While traditionally it was believed that mainly the large companies such as Yahoo!,[58] Canon[59] or Toshiba[60] are faced with multimillion-dollar patent infringement law suits, the new environment has also led to more aggressive enforcement against smaller infringers[61]. Lanjouw and Lerner (1996) for example found empirical support for the assertion that financially strong firms use preliminary injunctions to attack and oppress weaker firms by driving up their costs, substantially increasing their willingness to settle. Some 'patent trolls,' notorious patent holding companies that don't offer any commercial embodiment of the alleged invention but wait until someone infringes their patents to aggressively attempt to extract license fees,[62] such as PanIP for example, specifically sue small (high tech) companies who are less likely to be able to defend themselves (Davidson 2004).

The rise in patents has, especially in rapidly changing technology markets, made it increasingly difficult to innovate without infringing numerous patents.[63] In various industries, such as in semiconductors and biotechnology, there is such "an overlapping [and even conflicting] set of patents" that "those seeking to commercialise new technology [are forced to] obtain licenses from multiple patentees" (Shapiro 2001). Such a patent thicket is exacerbated by the rapid trend towards cross-disciplinary, multi-invention products.[64] The availability of patents in many of the involved fields has led to an "increase in [the] transactions costs [needed] to innovate."[65] According to Jaffe and Lerner (2004; p. 16) these developments "created a substantial 'innovation tax,'" as it effectively forces potential innovators to undertake complicated patent searches and to pay for numerous required licenses.

Many legal and economic commentators have expressed that patents can undermine the incentives of third parties to make improvements to patented technologies.[66] While the total amount of innovation does not seem to have decreased as a result of recent developments,[67] the changes seem to be especially burdensome for potential user-innovators. These generally obtain substantial amounts of the needed know-how and tools from internal sources and the public domain.[68] As the increased patenting has limited the public domain in terms of the newest and most advanced technologies, in various industries it has become more difficult and costly to acquire critical know-how.[69] In light of McGavock and Lasinski's (1998) survey, where of 186 Licensing Executive Society members only 3.4% of respondents stated that they would ignore minor infringers, the option of ignoring patents seems to often "no longer [be] an economically feasible option" (Kastriner 1991; p. 15) if user-innovators intend to diffuse the results of their tinkering, which they often do.[70] Williams (2005; pp. 35–36) thus states that increased patenting "risk[s] crowding out...open science." Due the lack of skills and experience to successfully navigate the patent thicket[71] and the involved financial risks[72] the mere threat of patent infringement litigation is often sufficient to stall users' innovative activities. Jaffe and Lerner (2004; p. 6) summarized somewhat cynically that "increasingly, the firm with the best lawyers or the greatest capacity to withstand the risk of litigation wins the innovations wars—rather than the company with the brightest scientists or most original, valuable ideas."

Users faced with such threats can either seize their activities or pay for a license.[73] Due to their limited finances litigation is often not an option.[74] While licensing may be a suitable option for users, it still may require a substantial investment in terms of time and money.[75] Patentees may furthermore refuse to grant a license.[76] While parties who own a strong portfolio of relevant patent "bargaining chips" (Cohen et al. 2002) may manage to negotiate a cross-licensing agreement this option will not generally be available to users.[77] As users generally improve and modify to meet their own needs and requirements[78] and due to the costs of patenting, especially internationally, they will frequently lack patents in a specific area.[79, 80] Von Hippel and von Krogh (2006; p. 300) summarize that patents are "especially impractical for many individual innovators, and also for small and medium-size firms of limited size." This will be especially pronounced for users in consumer goods sectors.

Not owning patents can have dramatic consequences. Not only are patents of critical importance in obtaining funding such as venture capital,[81] but they are also often necessary for survival in the market place. As Rivette and Kline (2000; p. 181) state in regard to innovation on the internet: "Patents can become your only shield—your only defence and proprietary advantage against bigger or better-financed competitors." Schiffer (2003) concurs and explains that patents enable small professional inventors to exist. Without patents however, users may struggle to obtain access to state-of-the-art technology in numerous field from pharmaceuticals to electronics.[82]

Even if users do possess IPRs in the field in which they would like to innovate, their relevant IP portfolio is still likely to be limited as innovation and patenting usually will not be their principal role in the industry's value chain. There are strong indications that parties with small IPR portfolios face considerable disadvantages in enforcing their patents and obtaining access to needed third-party patents. An EU commissioned study of 600 European small- and medium-sized enterprizes (SMEs) who had obtained patents between 1994 and 1997 found that 67% had experienced attempts at copying their patented inventions, but only around 20% actually used the courts to remedy the infringement and defend their patents.[83] Lanjouw and Schankerman (2004), in their study of 9,345 patents involved in litigation and settlements between 1978–1999, found that the probability of suit is inversely related to the size of a company's patent portfolio. In other words, the patent litigation risk is much higher for individuals and companies with small patent portfolios due to their inferiority in a) trading their intellectual property and b) in relying on repeated interaction with other patentees. These findings strongly suggest that there are barriers to innovate for those who lack a strong patent portfolio or cooperation potential in that specific field. This is also because inventing around a single patent will be much easier than circumventing a patent portfolio.

Users wanting to patent will face various further disadvantages as opposed to manufacturers. First of all users are often likely to freely reveal their developments to test whether there is a market demand that would make patenting worthwhile.[84] If however they have not applied for a patent one year after the disclosure, most patent laws will prevent them from being able to obtain a patent. Another factor that Shah (2000) reports in regard to several sports-related user-patents is that many users do not actually meet the standards for obtaining a patent—especially in terms of the "non-obviousness" requirement. Shah and Tripsas (2004; pp. 19–20) mention an example of a windsurfer who was granted a patent on "camber inducers" that are supposed to give the sails more shape. The innovator was able to strike a licensing deal, including the provision of consulting services. Although he initially received his royalties in good order, once the licensed manufacturer had received a sufficient amount of know-how, he ended the user's consulting work and stopped making royalty payments. While the dispute was eventually resolved in court, it illustrates the advantages of large firms over individual user-innovators. To enforce their rights the latter will often have to use the courts, which, given the financial and temporal costs, can be a considerable burden.

While the patent thicket can be mitigated in various ways neither a merger, due to the involved costs, nor the creation of a patent pool[85] is likely to be a feasible option for most users. The latter often creates "a barrier to entry into the industry" (Kaiser and Ronde 2005; p. 7) by frequently excluding new entrants. For instance, via cross-licenses between the leading players in the seed industry such as Monsanto, DuPont, Bayer or Dow AgroSciences a global technology cartel was created making innovation for seed-users practically impossible.[86] Similarly, the Summit/VISX patent pool monopolized the market for the lease and sale of medical equipment used by doctors to correct certain vision disorders. It thereby could squash user-innovation as the doctors using the system could not have diffused an improvement in this area without the consent of the patent pool.[87]

Finally, consider the beginnings of the motor industry. After purchasing a broad patent on the gasoline-powered internal combustion engine that was issued to George Seldon in 1895, the Electric Vehicle Company (EVC) demanded high royalties from automobile manufacturers. After years of expensive litigation these formed a patent pool with EVC, known as the Association of Licensed Automobile Manufacturers (ALAM), which severely restricted the group of auto manufacturers to whom the benefits of pool membership would be extended. George Day, founding member of ALAM and President of EVC stated that ALAM "will license no unreliable start-ups." Most manufacturers, who applied, including Henry Ford, were denied licenses. While this regime subsequently became much more open, after a policy reversal the attitude towards small companies that provided no patents to the pool became hostile again.[88] Rai (1999; p. 133) therefore generalizes that "entities may be denied access to the patents held by a pool if they do not have a new technology to contribute to the pool or are unable to pay large amounts of money to purchase a license."

Therefore the only remaining "working solutions" for users that don't substantially increase the costs of their innovation—inventing around patents, challenging patents in court or going offshore generally require the investment of substantial resources[89]—are direct infringement[90], often informally invoking uncertain research exemptions,[91] and the utilization of the limited amount of publicly available tools.[92]

To further illustrate our argument the booming field of nanotechnology[93] is enlightening. Practically all Fortune 500 companies are heavily investing in this field—over $50 billion worth of nanotech products were sold and $12 billion were invested globally in 2006[94]—which is likely to revolutionize the worlds' economies over the medium to long run.[95] As a wide variety of nanotech-patents have been issued,[96] various commentators agree that a patent thicket is slowly forming that is likely to make life increasingly difficult for user-innovators in this area.[97] This is especially as the basic concepts of nanotechnology are being patented.[98] Lemley (2005; p. 9) therefore argues that as with other "enabling technologies" such as computer hardware,[99] software,[100] the internet[101] and biotechnology,[102] "the basic building blocks of the field were unpatented," such intense patenting can pose a particularly strong threat to nanotechnology innovation. This is especially as their breadth[103] make designing around nanotechnology patents harder than in other fields.[104]

Anyone wishing to use the building blocks of nanotechnology "will be forced to license patents from many different sources in order to do so" (Matthew Nordan cited in Hebert 2005). This is also due to the cross-disciplinary and multi-industrial nature of nanotechnology.[105] As nanotechnology inventions can be kept secret relatively easily, the likelihood of users infringing patents in case of improvement innovations is also increased due to the difficulties of being aware of the existence of incorporated patented technologies in products.[106]

Obtaining licenses to nanotech patents will therefore be burdensome for users as they first have to identify patents from a wide range of industries which are potentially infringing and then attempt to obtain licenses, which patent holders may often be unwilling to grant to users.[107] This will be especially difficult as many patent holders prefer to license exclusively,[108] at least to cover a specific industry, as this leads to higher royalties. If licenses are available, the involved transaction costs[109] of having to negotiate multiple licenses from various parties is likely to limit user-innovation to exceptional cases, especially as users are unlikely to be as able to dedicate the required time and resources as those firms whose professional role it is to commercialize innovative nanotech products.

Another area in which numerous commentators agree that the law has been creating barriers to user-innovation is the area of software patents.[110] As the open-source movement is one of the best examples of user-innovation,[111] this section will discuss the patent barriers that such users face.

In the beginning of the software industry, during the 1970s, software was not patentable. This changed however, at least in the US, in the early 1980s.[112] Since then there has been a tremendous growth in software patents.[113] Alone in the US around 20,000 software patents are issued annually,[114] leading to a patent-thicket in this area.[115] In a study commissioned by the German Ministry for Economics and Technology 20% of the interviewed firms from the primary software sector and 40% from the firms in the secondary sector, i.e. manufacturing companies that also produce software, responded to have already been involved in some form of patent dispute.[116] Due to this immense increase of software patents since the early 1990s, innovators in the industry are increasingly faced with the threat of litigation.[117] This is especially as software development is generally a cumulative process, in that most programs are based on previous ones. Most open-source programs for instance fall into this category: Linux is based on Minix[118] and Firefox was created from the remains of Netscape's Internet Browser.[119] Creating programs from scratch is often too difficult, expensive and time-intensive. The increased patenting of software therefore has the potential to "enclose the digital commons" (Poynder 2003a; p. 37). In this way, if the existing patents are eventually enforced, open-source developers seem to have fewer opportunities at their disposal for their creative and innovative endeavors.[120]

While it will all depend on how these patents are eventually enforced and treated by the courts, many commentators are convinced that software patents threaten user-innovation. Tim Berners-Lee (2000; p. 213), the developer of HTML, states that software patents present a "great stumbling block for Web development." Henkel and von Hippel (2005; p. 74), as strong proponents of user-innovation, acknowledge that "there is good reason to assume that the current tendency towards stronger intellectual property protection has a negative impact on user-innovation." Richard Stallman (2002; p. 105), the notorious free software proponent, explains that "the chance you can get all the way through the [patent] minefield and develop the programme you want to develop without ever stepping on a patent gets less and less as the program[s] get bigger." The Linux operating system for instance allegedly violates 283 patents, although none has yet been asserted.[121] The fact that patents are written in legalistic language that is often difficult to fully understand for non-lawyers contributes to the difficulty of navigating the IP thicket for users. How could the developer of a natural-order recalculation spreadsheet program have discovered the 1964 topological algorithm patent that could have blocked his software?[122] Further examples of software user-innovations that had to be terminated due to the existence of patents are the predefining abbreviations feature of XyWord, an early MS DOS and Windows word processor; the compress program, written in 1984 and blocked by a patent on an algorithm issued in 1985,[123] and the experience of Vern Blanchard. He wrote a software program in 1993 for the use in professional bingo halls which he started to market when a competitor filed suit for patent infringement. Although Blanchard eventually won in court, as his competitor had won an initial court injunction that prevented him from bringing his products to the market, as well as scaring off prospective clients, Blanchard was put out of business with debts of $100,000.[124] Henkel (2002) generalizes this to state that open-source programmers will often lack the ability to arm themselves against patent infringement suits by commercial competitors.

A further problem of the US patent system[125] for the software industry is the one of continuation or submarine patents. In such cases the patent application is re-filed, sometimes various times, to postpone the application process and adapt the application to current developments. While the law has recently been changed to constrain such practices,[126] continuations remain a problem in the software industry.[127] Graham and Mowery (2004) believe that the "continuation practice offers a particularly effective hold-up instrument when used against software innovators because returns to investment in software often depend on being first-to-market and the creation of a critical mass of adopters." They argue that this is especially dangerous for the open-source movement. To illustrate their arguments they cite the *Rambus* case[128] where, after various re-filings, a patent was issued to Rambus 9 years after the first application was made. Rambus subsequently started to demand royalties of up to 3.5% on sales of the major DRAM manufacturers worldwide. This was possible as the patent was amended to exactly fit the communication standard that the semiconductor industry had adopted in 2000. Hitachi, Mitsubishi, NEC, Samsung and Toshiba were reluctant to risk expensive patent litigation and agreed to pay royalties. Infineon, Hyundai Electronics and Micron Technology on the other hand took Rambus to court. As the CAFC rejected the challengers claims that Rambus had misused the continuation procedure by altering claims in a pending application, by amending them to cover new uses for a technology being developed by competitors or customers, Graham and Mowery (2004) conclude that "after *Rambus*, the law appears to support an applicant filing an early, broad patent application on a software innovation and then watching the open-source code develop." While big semiconductor corporations can afford to litigate or to license depending on their desires, small sized companies or individual user-innovators are likely to be in a worse bargaining position.

Another example of the power of patents in blocking user-innovation is the case of public key encryption, which is an essential feature inter alia of the internet, electronic banking and data warehousing.[129] Until the expiry of the main patent in 1997 nearly all innovation in this area were blocked as the patent owners "crushed" numerous programs that were created (Stallman 2002; p. 99) and as the patent's breadth practically prevented inventing around it.[130]

At least two types of user-innovation communities have been threatened with patent infringement litigation: the WAP-Forum and the Java community. A patent battle between the two holders of the most critical WAP patents temporarily endangered the innovative efforts of the WAP Forum, before these finally settled and allowed users to continue to develop WAP applications.[131] The authorized Java community, that developed applications for Sun Microsystems' programming language, was threatened by Eastman Kodak's patent infringement suit against Sun in early 2002. As in October 2004 a federal jury in New York upheld Kodak's suit to find that Sun had infringed three Kodak patents, these communities were faced with having to seize any further development. Although Kodak claimed over $1 billion in damages, fortunately for Sun and its users, the parties subsequently settled, with Sun and its communities obtaining a right to use the relevant patents.[132] Without a powerful company to protect such user-innovation communities, these may not have been able to continue their innovative tinkering.

To summarise, while our examples were anecdotal, various empirical studies support these indications. Besson and Hunt (2004)'s and Cohen and Lemley (2001)'s studies indicate that software patents burden innovation by encouraging strategic patenting. Blind et al. (2003; pp. 8–9) found that companies with an open-source business model fear that the extension of patenting "will have without exception negative consequences." None of these studies are conclusive however and further investigations are needed in this controversial matter. Overall, much will depend upon how the issued patents will be enforced and whether they will be upheld by the courts. Enforcement is not per se a legal question, but rather a strategic and social one.[133] Whilst financial and legal reasons such as enhancing and protecting one's profitability may point towards enforcement, adverse publicity and customer outrage may point against it. It may therefore be worthwhile to analyse whether or not there is a business norm of some sort against suing innovating users. Despite of the Aibo case (below) or the IBM patent donations[134] we argue that the suits against individual music file-sharers and downloaders indicate that no such norm currently exists. While open-source user-innovation will nevertheless continue in the foreseeable future, the spread of software patents has certainly not facilitated such programming. As it seems equally unlikely that software companies will give up their IP or that the open-source software movement will completely halt their developmental activities, the future will increasingly be characterized by hybrid business models which combine the requisite strengths of the two.

## Copyright Law

Copyright also has the potential to adversely affect the ability to tinker. Copyright is a form of intellectual property that covers creative works such as most importantly music, art, literature and software. It protects the literary, musical, pictorial or graphical expression of a work rather than the underlying product itself.[135] Traditionally, copyright law according to the US Constitution and the relevant case law was intended to strike a balance between the protection of authors and the fostering of creativity.[136] According to many commentators however, recent legal and judicial developments have tilted the balance towards author protection.[137] Poynder (2003b) even argues that copyright, not patents, poses the greater threat to the digital commons. Lawrence Lessig (2001; p. 253) believes that digital copyright laws have given copyright holders "the power to control innovation in the context of the net." Michael Perelman (2002) similarly argues that new IP laws are helping corporations to confiscate creativity.

The effect has been especially pronounced in the realm of software. This is as copyright can protect individual lines of code, which cannot be reproduced in an identical manner. Due to the long duration of copyright, in the US it now lasts for 70 years after the author's death, this can provide substantial difficulties for those wanting to tinker with software by reusing exact lines of code, which is a frequent practice to cope with complexity.

To exemplify how copyright law has been used to prevent user-tinkering, the story of Jesse Jordan is useful.[138] This computer-science student of Rensselaer Polytechnic Institute developed an intranet search-engine by tinkering with other search engines. His main improvement was to fix a bug in Microsoft's file-sharing system that could cause a computer to crash. His search engine, from which he did not intend to derive any profit, went on-line in late October 2002 and after several amendments it was running smoothly by March 2003. One month later Jesse was informed that the Recording Industry Association of America (RIAA) was filing a lawsuit against him demanding $15 million due to facilitating illegal file-sharing. As the costs of defending his case would have been at least $250,000 he was forced to accept a draconian settlement where he paid the RIAA his life time saving of $12,000.[139] This example is just one of many cases—allegedly of already over 4500—in which the RIAA has sued private individuals in the US for either tinkering with peer-to-peer (P2P) technology or for illegitimately downloading music.[140]

While litigation per se can adversely affect innovation in an area in the short run, it often leads to more legal certainty in the longer run. The many lawsuits in this realm, especially in the US, have not yet had this effect. Lessig (2004; p. 192) therefore calls the current law "a mess of uncertainty . . . If innovation is constantly checked by this uncertain and unlimited liability, we will have much less vibrant innovation and much less creativity." While the breadth of copyright protection and the extent of its effect on user-innovation will vary from country to country and have so far fortunately not halted open-source innovation, the strength of copyright will force many software users to tinker with non-copyrighted open-source software.

## Anti-hacking laws

In recent years anti-hacking legislation and enforcement mechanisms have significantly increased globally. Not only have the scope of such legislation and the severity of punishment increased, but also has the likelihood of getting caught. These changes threaten to deter a wide variety of hackers from tinkering.[141]

The most notorious of these is the US's Digital Millennium Copyright Act (DMCA). This Act was merely intended to implement WIPO sponsored international copyright treatises which adapt the law to the evolution in technology that allows the nearly costless and immediate duplication of copyrighted material. To attain this objective, anti-circumvention provisions were included in the DMCA: "no person shall circumvent a technological measure that effectively controls access to a [copyrighted] work."[142]

This covers the descrambling of a scrambled work, the decryption of an encrypted work or otherwise avoiding, bypassing, removing, deactivating or impairing a technological measure without permission of the copyright owner.[143] Those who fail to abide to its stringent rules can face both civil and criminal liability.[144]

While originally it was intended to only cover copyrighted digital content, the DMCA has had a far wider actual effect, from encryption research to journalism and even to hardware.[145] This wide effect is especially due to the absence of fair-use exceptions, such as for non-commercial tinkering or making back-up copies. Although there are exemptions, these have generally been considered to be insufficient. Samuelson (1999; p. 19) explained that only three exemptions "specifically authorize the creation of tools necessary to achieving legitimate circumvention ability": the encryption research, computer security, and interoperability privileges. The latter allows those who "lawfully obtained the right to use a copy of a computer program" to circumvent a technological copy-protection measure "for the sole purpose of identifying and analyzing those elements of the programme that are necessary to achieve interoperability of an independently created computer program."[146] As this exemption was narrowly interpreted in *Blizzard* it is unlikely to significantly decrease the breadth of the DMCA's prohibition. Other commentators also referred to the exemptions as "so parsimonious to be of little practical value" (Bellovin et al. 2001).

The DMCA even turns the previous law of generally allowing reverse engineering subject to exceptions upside down to prohibit such practice except in limited circumstances. In other words, although the Act's declared purpose was to balance the interests of "content owners . . . and information users in a way that will foster the continued development of electronic commerce and the growth of the internet," its implementation seems to protect the interests of media corporations.[147]

Other nations, most notably Australia and Japan, have followed the restrictive precedent of the DMCA in adapting their copyright laws.[148] The European Union's Copyright Directive (EUCD)[149] and its national implementations seem "uncannily similar" to the DMCA, as English law firm Wragge & Co. (2004) emphasized. Knowingly circumventing technological protection will for instance for the first time constitute a copyright infringement in many European countries.[150] If undertaken with a business motive it will be a crime.[151] Article 6(2) of the Directive in very broad terms requires EU members to provide "adequate legal protection against the manufacture, import or sale of devices which are promoted or designed primarily for the purpose of circumventing an effective technical measure." Commentators have furthermore pointed to the inadequacies of EUCD's consumer protection mechanisms.[152] Dusollier (2003; pp. 54–55) therefore stated that it contributed to the "shift of copyright from public law to a regime of private orderings enabled by both contract and technological means."

Hacking, for our purposes creating "code that enables the programme to do something it was not originally intended or enabled to do" (Lessig

2004; p. 154), has been criminalized in most nations.[153] Simultaneously, the penalties for cybercrimes have become much more draconian[154] and the establishment of anti-cybercrime agencies[155] has resulted in more effective pursuits and crackdowns against cyber-criminals.[156]

Due to their breadth[157] such laws generally cover both illegitimate and legitimate hackers. In other words, such laws can burden those who tinker in order to develop new products or to enhance security. This is due to the difficulty for tinkerers in this realm to know whether their activities are legal or not, even in regard to modifying legitimately acquired software[158] (cf. contractual barriers). Those tinkerers who like to see themselves as law-abiding may well be deterred from their activities not just due to the fear of being caught but also due to the social stigma attached to such grey activities.

As renowned security analyst and hacker Marc Loveless states: "[a] lot of things we [hackers] used for research [purposes] ... can now be considered a criminal act" (Hulme 2003a). Indeed, even those who simply write hacking software for research purposes or study software to discover security holes have been faced with liability.[159] This is especially as the the DMCA has made tinkering a crime if it involves the circumvention of technological barriers involving copyright. For instance, a team of researchers led by Princeton Professor Edward Felten successfully responded to a Secure Digital Music Initiator (SDMI) public challenge encouraging skilled technologists to defeat certain watermarking technologies. When the researchers wanted to present their results at a conference, SDMI and RIAA threatened the researcher with liability under the DMCA, leading to the withdrawal of the paper. Although they subsequently published their findings, they first had to obtain a court declaration that they had First Amendment Rights to do so.[160] Samuelson (2003; p. 42) consequently stated that the DMCA "impedes the progress of science" and that it already has had a "chilling effect" on the publishing of research of cryptographers. Indeed, a lot of information that was once openly shared on the web is now increasingly available through invitation-only groups or encrypted mailing lists.

The EFF (2003; p. 1) concluded that the DMCA and its anti-circumvention provisions "have been used to stifle a wide array of legitimate activities, rather than to stop copyright piracy": It has been used to "chill free expression and scientific research," to "jeopardize fair-use" rights and to "impede competition and innovation" (ibid.). To start with some examples, let us consider the case of censorware researcher Seth Finkelstein. Censorware blocks websites containing objectionable material. As such programs generally contain technological access controls to protect the lists of blocked websites researchers in this area may breach the DMCA by circumventing such access controls. Finkelstein therefore only conducted his research due to a limited 3 year exemption that he received from the Librarian of Congress and Copyright Register in 2000. However as the librarian has no authority to limit the anti-trafficking provisions, even though Finkelstein can legally circumvent technological measures, he cannot distribute his (innovative) circumvention tools, or use his findings to develop innovative products.[161]

While users may continue to develop promising products, they are less likely to be distributed or commercialized. Take for instance Other World Computing (OWC), a small Apple Macintosh retailer that developed a software patch which allowed all Mac owners to use Apple's movie saving software iDVD. This software was designed to only work on newer Macs containing Apple's internal DVD recorders. OWC's minor software modification allowed iDVD to also work with external DVD recorders, giving owners of old Macs an upgrade possibility. When OWC started to sell its software, Apple's threat of DMCA liability practically forced OWC to withdraw its development immediately from the market.[162] EFF (2003: p. 12) therefore stated that "rather than prevent copyright infringement, the DMCA empowered Apple to force consumers to buy new Mac computers."

In the *Reimerdes* case the court supported these views. The defendants had purchased DVDs, which due to the lack of an authorized DVD player application for Linux, they were unable to play. They had then used DeCSS to play their DVDs. The court held them to have violated the DMCA as they circumvented CSS without authorization.[163]

A final example is the case of the Russian programmer Dimitri Skylarov, who wrote the "Advanced e-book processor" software for his employer. The program enabled users of Adobe's ebook Reader to disable restrictions that an electronic book publisher may have imposed.[164] Although the program was legal in Russia, Adobe Corporation complained to the FBI, which arrested Skylarov when he came to the US to attend a conference. Before being entitled to go home, he was jailed for various weeks and detained in the US for 5 months. The adverse publicity of these cases "has appeared to curtail the legitimate research activities of computer security researchers" in the US (Tang 2005; p. 868). Indeed, various other prominent computer security experts have either curtailed their research activities or halted the publishing thereof.[165] While "the jury is still out on whether the pace of [development of] decryption technologies will indeed continue to slow down or be banned altogether" (Tang 2005; p. 869), it seems that the DMCA adversely affected user-innovation in this realm, at least in the US.

While some hackers may be resilient to criminal sanctions,[166] many sensible hackers are likely to contemplate halting their work due to the fear of litigation.[167] In a survey of over 2,000 CEOs, 47% of manufacturers reported to have withdrawn products out of fear of litigation.[168] If manufacturers suffer from such legal barriers, users are unlikely not to be detrimentally affected.

## Why IPRs are Likely to Remain Strong

We have seen that strong IPRs can have a detrimental effect on user-innovation. We will now argue that there is considerable evidence for the assertion that globally IPRs will remain strong in the foreseeable future.

To start with we should briefly consider the four main reasons for the emergence of the pro-patent era in the early 1980s:[169]

- a change of the political attitude, under the Reagan administration, which started to combat the "free-riding of US technology" by Asian economies by demanding stronger global protection of Intellectual Property;
- the adoption of a significantly more liberal approach to IPRs and IP licensing in regard to anti-trust matters by the US Department of Justice;
- fierce demands by US big industry for stronger IP protection and enforcement against infringers and counterfeits domestically and abroad; and
- the establishment of the CAFC in 1982 with its pro-patent attitude.

All these factors caused the economic value of patents and other intellectual assets in the US and abroad to rise significantly. For example, a 1997 study by Coopers & Lybrand (1997) found that two-thirds of the then $7 trillion market value of all publicly traded US companies lies in their intangible assets such as their brand equity and their intellectual property. In the *Unseen Wealth* report of the Brookings Institution (Blair and Wallman 2001) it is further argued that intellectual capital, R&D, brand names and human capital have become the primary drivers of economic growth. While in 1982, the Brookings Institution reported that 38% of corporate market capitalization could be attributed to intangible assets, 10 years later already 68% were considered intangible. In 1998 it reported that 85% of the value of the S&P 500 was attributable to intangible assets.[170] Granstrand (1999) therefore stated that we have entered the age of "Intellectual Capitalism," where productive assets and commercial transactions are predominantly intellectual rather than physical in nature.

We strongly agree with Josh Lerner's (2004) argument that the changes that initiated the pro-patent era, especially the higher probability of successfully litigating and the expectations of obtaining greater awards, have created a vicious (or virtuous) cycle (depending on one's view). Once one firm starts to aggressively enforce its patents and succeeds in doing so, its rivals are bound to follow. We indicated how companies such as Texas Instruments opportunistically enforced their IP, generating tremendous sums with relatively little effort. The almost inevitable result was a "quasi arms-race" between competitors in the industry, which initially rapidly spread throughout the US economy to then subsequently proliferate internationally (Chartrand 2004).

The Danish Wind Turbine industry offers a good example. Free sharing of the original designs and improvements were the foundations of the Danish success in this realm. Patents played a minor role at most. In 1998 while Danish companies held over 50% of the world market, they filed only

1.5% of global wind energy patent applications.[171] Due to competitive pressure from the US, especially due to the increasing amount of patents being granted on novel types of wind turbine designs,[172] the Danish free-revealing culture came to an abrupt end and Danish companies have been forced to also focus on patents. World-market leader Vestas (2005; p. 51) therefore stated in its Annual Report 2005 that while it "previously failed to devote sufficient focus to intellectual property rights . . . in future [its] goal is . . . to increase the number of patent applications significantly." More precisely it realized that "only through strong patents will Vestas be able to protect its future earnings capability" (Vestas 2005; p. 44).

Given the financial value and strategic importance of IP to patent owning corporations, Benkler (2002; p. 95) argues that this expansion of IP protection will create "feedback effects that amplify the direction and speed of the shift in strategies, and lock them in institutionally." Granstrand (1999; pp. 41, 323 and 326) similarly argues that

> in the late 1990s there were no signs of a reversal of the pro-patent era, on the contrary... [most patent owning] corporations will not easily accept a devaluation of their immaterial assets. Rather they will lobby for a continued strong patent regime.[173]

It is rational behavior to invest in sustaining favorable institutional arrangements.[174] An enlightening example is how US universities reacted to proposed changes to the Bayh-Dole Act, which gave them the right to exploit government-funded research through patents. As universities were now able to generate considerable more income from patenting their findings[175] they started to actively oppose changes to this policy, despite of the alleged detrimental consequences for free R&D.[176] Recently Charles Phelps (2005), on behalf of the Association of American Universities, American Council of Education, Association of American Medical Colleges and Council on Governmental Relations, explained that all of these organizations would "oppose statutory changes that would weaken patent rights."

Corporations, which focus to a much greater degree on their bottom line than universities, are likely to be even more reluctant to give up a part of their wealth. Rising R&D costs[177] with receding returns,[178] the increase in technology-based competition,[179] constantly increasing IP theft[180] and generally no better alternative to encourage innovation,[181] will ensure that the pressure will remain high for strong IPRs for the foreseeable future.

To illustrate the influence of powerful media, pharmaceutical and technology companies on IPRs we should quickly refer to the CTEA. While previously copyright protection already lasted for 50 years after the author's death, the CTEA, without an apparent need, enlarged this by a further 20 years.[182] Although the official purpose of the Act was to "stimulate the creation of new works and provide enhanced economic incentives to preserve existing works,"[183] this has been harshly criticized as being pretence.[184] Rather than stimulating the creation of more creative works, it does seem that the additional 20 years of protection after the creator's death merely give existing copyright holders more time to exploit their rights.

Various other lobby groups favor stronger IPRs.[185] The best examples are the Motion Picture Association of America and the RIAA, who both heavily lobbied for the CTEA and the DMCA and will continue to strongly advocate the strengthening of copyright in the future. The fact that both are still dissatisfied with the status quo can be illustrated by the thousands of law suits that they have launched against private individuals[186] and further attempts to introduce restrictive legislation.[187] In fact, to combat the rising global tide of unauthorized copying and usage of digital files, many media corporations have started a "counterrevolution with all technical and legal means" (Staun 2006).

All this is not intended to be overly one-sided. There are counter movements such as the Electronic Frontier Foundation (EFF), which has over 50,000 members which are actively fighting to restore the balance of copyright law.[188] While this movement has led to various promising suggestions[189] that were considered to finely balance the interests of the media industry with those of consumers,[190] their reform efforts have, largely due to the strong majority with which the DMCA originally passed in Congress, and the lobbying power of media conglomerates in Washington, not yet been fruitful.

Global pressure, especially on developing countries, to create, preserve and enforce strong IPRs is likely to remain intense. Pamela Samuelson (2004), Law Professor at Berkeley, discusses a concept called Intellectual Property Arbitrage. As products entering a local market from another country with weaker IPRs can undermine the force of IPRs in that local economy, the pressure on such other countries to increase their IP protection will remain intense.[191] The lack of adequate protection against "parallel imports" for example is one of the main reasons why Western companies are dissatisfied with TRIPs and will continue to lobby for its strict implementation and enforcement.[192] Alone the threat of large MNCs to withdraw their investments from a country, which can be worth billions of dollars, has the potential to influence national IPR policy.[193]

Especially the US will continue to pursue its policy of pressurizing other nations to change their IP policies to suit its needs.[194] For this purpose the US Trade Representative (USTR) investigates whether particular nations adequately protect IP rights, and if not it is authorized to deny such countries trade benefits.[195] The USTR in 1997 for example withdrew $260 million of benefits to Argentina due to the allegedly inadequate manner in which it changed its trade laws.[196] Recently, the US has exerted pressure on China to strengthen its enforcement mechanisms[197] and persuaded numerous countries, such as Chile, Australia and Singapore, to adopt TRIPs-plus regulations and tougher enforcement standards.[198]

If we consider the other factors that initiated the pro-IPR age, we similarly see little hope for a weakening of IP. While the percentage of patents upheld in US courts has decreased slightly in recent years, this is mainly due to the substantial increases in patenting, including the granting of many bogus patents, and not a reversal of the pro-patent attitude.[199] Patent damages awards have similarly remained high,[200] the US Antitrust attitude remains liberal[201] and the US administration continues to favor strong IPRs. The EU seems to hold a similar opinion that strong and harmonized IPRs are desirable.[202] Via the EU Enforcement Directive[203] the rights of IPR holders were further enhanced by requiring Member States to make available preliminary injunctions, the seizure and recalling of infringing goods, and permanent injunctions.[204]

To conclude, we agree with Sell's (2003; p. 175) argument that "this broader IP regime is embedded in the deeper structures of global [knowledge-based] capitalism and American [and Western] power."[205] Until American commitments change or the connections between the broader IP regime and the structurally powerful are loosened, little is likely to change. Due to the inevitable gradual move towards "a borderless marketplace, [which]. . . demands that the territoriality of patent rights must also be broadened" (Blair and Wallman 2001; p. 75), we are likely to see a slow but steady trend towards the global harmonization of patent laws, especially once China shifts its emphasis from imitation to the creation of indigenous technologies. While an 'international patent' is still far away, regional harmonisation, such as in NAFTA or in the EU, are feasible in the short term. We have seen the potential of the IP expansion to adversely affect user-innovation and hope that lawmakers take the interests of users into consideration when they modify and re-tune their IP systems.

## MARKET BARRIERS TO USER-INNOVATION

The legal right to innovate is one thing; the ability to do so, another. In this section we will discuss both of these from an economic perspective: how companies use restrictive contracts, standards and their market power to exert control over users and their potentially innovative activities. Finally, we will briefly mention some organizational barriers to user-innovation.

### Contractual Barriers

To prevent user-innovation corporations can employ a variety of tools. The utilization of contracts is one of these. A basic example are breach of warranty clauses. If a user opens his computer, a prerequisite to improving the hardware, he will frequently be in breach of warranty, meaning that he could no longer return the computer if it malfunctioned during the warranty period. Dell Computers for example, informs customers ordering a new computer that opening the machine during the first 12 months voids the warranty. To ensure compliance with this, they attach foil stickers on the inside, which allow Dell to verify whether the machine has previously been opened. Similar practices are common in numerous industries from medical devices to printer cartridges to allow companies to protect themselves from liability for malfunctioning products. Even car manufacturers void their warranties when customers modify their vehicles by for instance experimenting with novel fuel-injection and ignition systems.[206] While this may not prevent users from tinkering, it does increase the costs of doing so. Anecdotal evidence indicates that some users, especially of more expensive products, are indeed discouraged from tinkering during the duration of the warranty.[207]

A more important tool to control how their technology is employed are modification prohibitions. Farmers for example, as we will explore more in depth below, are frequently faced with restrictive technology licenses which prohibit them from experimenting with proprietary seeds. Breaching these "gagging contracts"[208] leads to the risk of legal liability. In *Monsanto v Ralph*[209] a farmer who breached such a license was ordered to pay $3 million. While in theory users could switch to companies with less restrictive contracts, this may not be possible in concentrated markets where such licenses have become the industry standard.

Two further areas where restrictive End-User License Agreements[210] (EULAs) have become "ubiquitous" are the software and consumer electronics industries (Newitz 2005). EULAs have contractually prohibited millions of users from amending or modifying software programs.[211] A typical clause states that users "may not, in whole or in part, copy, photocopy, reproduce, translate, reverse engineer, derive source code, modify, disassemble, decompile, create derivative works based on the Program. . .without the prior consent, in writing of [the developer]."[212]

The courts have generally upheld EULAs.[213] Anti-reverse engineering clauses, which have been widespread from the early days of the computer software industry,[214] have however been more problematic, as they may conflict with the fair-use provisions of the US Copyright Act.[215] In *Bowers v Baystate Tech*, however it was held that "private parties are free to contractually forego the limited ability to reverse engineer a software product under the exemptions of the Copyright Act."[216] Consequently, a non-reverse engineering clause prevented a user-innovator, Harold Bowers, from improving a computer aided design program.[217] This holding was confirmed in *Blizzard*, where a US Federal Appeals Court in late 2004 held that a group of gamers, who had reverse engineered Blizzard's Battle.net software and developed a compatible new open-source gaming platform called bnetd, were in breach of a click-wrap license prohibition on reverse engineering.[218] In the EU reverse engineering can only be forbidden for non-interoperability purposes.[219]

As an increasing amount of software is patented[220] and as patent law contains no fair-use for reverse engineering, patentees can use contracts to prevent any undesired use of the whole program.[221] Indeed, this was part of the argument in *Bowers*. Sony also sued a playstation game emulator, Connectix, for patent infringement due to reverse engineering its technology.[222] In *Arizona v Lexmark*[223] such practice was upheld: "A restriction on a patented good is permissible as long as it is found to be reasonably within the patent grant." Although many patented goods are currently not sold with restrictive licenses,[224] the increasing prevalence of software and embedded computers in mobile telephones, cars, washing machines or medical systems[225] and the ongoing technological convergence (below), are likely to lead to a change of practice (as was the case with seed-purchasing farmers). Anecdotal evidence already indicates that purchasers of embedded products have to agree to technology licenses that forbid the modification of such devices.[226]

If such restrictive terms can be enforced it can be argued that control over information and software will shift to content owners and software producers.[227] It could adversely affect user-innovation.[228] Samuelson (2002; p. 18) expressed that enforcing non-reverse engineering clauses threatens "the continued competitiveness and innovation in the California computer industry—and indeed, of all industries that could impose mass market restrictions on reverse engineering." She bases her argument on the US Supreme Court, which considered reverse engineering to be an "essential part of innovation" that "could lead to significant advances in technology."[229] Stallman (2002) adds that such restrictions will result in a) fewer users of such programs, b) users being unable to adapt or fix them and c) developers not being able to learn from or base new work on them.[230]

Even if restrictive EULAs are not legally enforceable, the high legal uncertainty in this area,[231] which has been described as "the enemy of investment" (Jaffe and Lerner 2004; p. 194), and the negative publicity of the enforcement of such licenses in the past (e.g. in the *Blizzard* case), are likely to have a detrimental effect on user-innovation.[232]

Single-use restrictions, which the courts have generally upheld, are a further contractual tool that allows manufacturers to specify that a product can be used solely for its designated purpose. Microsoft for instance attempts to limit the use of its software to one computer,[233] and Lexmark tries to restrict users to only utilize printer cartridges once (*Arizona v Lexmark*). If the product is used for another purpose users can be liable for breach of contract and/or patent infringement.[234]

Overall, the usage of such clauses, especially if reinforced by legislation or the courts, is likely to lead many users to obtain "the impression that most kinds of tinkering are unlawful" (Newitz 2005). This could burden their creative and explorative activities.

## Market Power, Standards and Platforms

Another key barrier to user-innovation is the exercise of market power. Although there is little empirical support for a rise in corporate power in recent decades,[235] the available evidence indicates that corporations remain powerful[236] and that the perception of corporate power has increased.[237] In the following we will argue that companies often use their market power (including their IPRs) by inter alia bundling services and products, attempting to influence related markets, and by implementing standards and platforms[238] to control and increase their rents.

Consider the exercise of market power by the motion picture industry. Using their joint patent rights they forced DVD manufacturers to agree to the terms of the Content Scrambling System (CSS) license.[239] This license inter alia obliges producers to incorporate copy control signals, to refuse to play disks containing unencrypted content where content marks indicate that the material originally was encrypted using CSS (§ 6.2.13.2), to ensure that the regional playback controls are closely coupled with the authentication process to prevent any attempts to reengineer them (§ 6.2.2.2) and to implement future copy-control standards (§ 6.2.8.2). Gillespie (2004; p. 245) explained that

> this technologically enforced licensing scheme is the most egregious intervention into the distribution of cultural expression: By locking up their DVDs, the movie studios can ask almost anything of the hardware manufacturers, who can only obtain the key (necessary to even produce a DVD player) if they agree to every demand.

In other words such control enables a form of "innovation lock-out" as Lockton (2005; p. 21) phrases it.

In an age in which we are increasingly dependent on complex technologies, coordinating architectures, such as platforms[240] and standards,[241] are crucial. As complementary products are necessary for the success of a platform,[242] even market leaders have become dependent on others to be able to profitably innovate.[243] High-tech markets will therefore often be characterized by an intricate web of competition and co-operation between platform creators and producers of complements.[244]

This development could lead to an "increasing ability to innovate by more actors," including users, as platforms can lower the costs of coordinating innovative developments (Gawer and Cusumano 2002; pp. 3 and 6). As the existence of a dominating platform or standard in an area often means that subsequent innovation has to be complementary to such standards or platforms, infrastructure providers, such as platform-leaders or Standard-Setting Organizations[245] (SSO) can frequently exert a considerable amount of control over third party innovation and its costs. This is because non-compatible products have to be substantially better than current products in order to lure customers away.[246]

The extent of such control will depend on whether purely proprietary or completely open standards/platforms will prevail. Harvard Law Professor Jonathan Zittrain (2005; pp. 18–19), for example argues in favor of the former: "Many emerging video game, cell phone [Apple's iPhone], and PDA platforms [Apple iPod or Microsoft's Zune] are closed from the start—third-party developers either are not welcome or are subject to stringent licensing requirements." In closed platforms related IPRs will be closely controlled[247] and the technical interface specifications will not be publicly available. In this way, third parties, such as users, will find it "very difficult—if not impossible—to design complements."[248]

The new generation of Japanese mobile phones, the increasing merging of phones with PDAs, further indicates where the trend may be going. Content in these gadgets is generally locked to the cell phone it was downloaded with. Any kind of reproduction is not allowed and backups, at least in theory, are not possible. Bovens (2005) fears that such high level of control over content will have a "chilling effect on future creations." This will make the creation of complementary applications or an interoperable platform depended on obtaining permission from the platform controller.

Sanchez (2002) on the other hand, believes that there is a general shift from proprietary to open and modular architectures,[249] in which the technical interface specifications are publicly available to encourage third party innovation. Open and modular product architectures divide complex design tasks into smaller modules, which allow user-modifications.[250]

The truth, as often, probably lies somewhere between these two extremes, as corporate success in this realm will depend on finding an adequate balance between controlling and encouraging complementary innovation.[251] Lemley (2002) therefore describes that most telecommunications, semiconductor and software standards are neither open nor proprietary, but instead occupy a middle ground, where anyone can join, who is willing and able to pay some form of royalty to the IP owner(s).[252]

Furthermore, although most open standards are based on the RaND principles (i.e. reasonable and non-discriminatory licensing commitments) they nevertheless often contain restrictive clauses such as field-of-use limitations and reciprocity requirements, such as having to license requisite patents to the standard. Intel for instance used its open PCI and USB standards to prolong its dominance of the PC hardware[253]. Giving up one's IP rights in an area may be especially burdensome for individual innovators or small companies.

The main problem of an open-non-proprietary architecture is appropriation by competitors and loss of control over its subsequent developments.[254] Take the development of the personal computer, developed by IBM. While its open-architecture originally led to a quick market penetration, it also meant that IBM lost control over the technology and allowed Intel and Microsoft to respectively take over the microprocessor architecture and operating system.[255]

Even open standards can be "a powerful way to raise barriers to entry" (Cusumano and Gawer 2002; p. 141). Although the interface to Cisco's IOS, the standard software for network integration, is open, Cisco keeps complete control over its software code and only allows its licensees to develop complementary products.[256] The DVD standard, created by various major electronics companies,[257] is another example. While this allows anyone to participate, the high royalty rates, which according to some estimates amount to nearly 30% of a DVD's average factory price, will burden many user-innovators.[258] In regard to proprietary genes, large pharmaceutical companies have created gene-platforms, which for many users, such as small biotech companies and public sector researchers, will be impossible to access due to their high costs.[259] Although some of these may be free for academics or independent researchers, if they want to commercialize their results heavy fees will be levied.[260]

Once a standard or a "dominant design" emerges, Abernathy and Utterback (1978) argue that products will become quasi commodities and their differentation in terms of function and feature will vanish. This will lead to a shift in emphasis away from product towards process improvements.[261] In other words, user-innovation would have to change form. It will however often be difficult for users to improve an established manufacturing or commercialization process. While some large corporate users may be able to develop process innovations such as in the realm of information technology—American Airlines commercialized its formerly proprietary reservation system Sabre to allow it to become the industry standard[262]—most end-users, especially consumers, will face considerable difficulties in this task. They will generally lack both the insight into manufacturers' internal processes and the consent from the key market players to modify the dominant design.[263] Dahlin et al. (2004) found that in the tennis racket industry firm-based inventors are much more likely to file patents concerning process innovations as are independent innovators. While users of scientific instruments in a study by Riggs and von Hippel (1994) were responsible for 82% of the innovations with regard to "new functional capability," they only generated 13% of innovations with regard to "convenience and reliability." We therefore argue that there are some indications for the potential of a dominant-design to adversely affect user-innovation. Nevertheless users may be able to find a new innovation niche in which no dominant design yet exists—as was the case with the Rodeo Kayak.[264]

As a further illustration, take the wind turbine industry. While Danish tinkerers initially shaped innovation in this area, today progress is dominated by large corporations. Eize de Vries, an expert in the wind energy industry explains that,

> the wind turbine industry started by building 'advanced blacksmith' designs that users could modify. Now manufacturers are building 2 to 2.5 MW machines with rotor diameters of up to 80m. No one attempts to modify these. (Douthwaite 2002; p. 91).

The technology has become too sophisticated, complex and expensive.

Baldwin et al. (2006; p. 22) summarize this with their model of how user-innovations are commercialized and change over time:

> First, one or more users of some good recognize a new set of design possibilities—a so-called 'design space'—and begin to explore it. . .at some point in this process, user-purchasers will emerge, interested in buying copies of user-innovations rather than making them for themselves. In response, some user-innovators may become user-manufacturers, using high-variable cost, low-capital methods to satisfy the demands of fellow community members. Later, as the no-longer new design space begins to be 'mined out' [i.e. the technology matures], better solutions become harder to find and the rate of user-innovation slows. As a result, investments in manufacturing methods involving higher capital costs and lower variable costs become feasible. . .User-innovators who require the very best performance continue to innovate however [although this has become more difficult], and also continue to sell smaller numbers of high-performance products to user-purchasers who are willing to pay for advanced designs. But progressive mining out of the design space leads to steadily lower rates of user-innovation, less entry, and eventually to the collapse of the user-driven experimental sector of the industry.

Another critical reason why dominant designs and standards can be barriers for user-innovation is because they can lock in users due to the involved switching costs.[265] These are recognized to significantly affect user-behavior in a wide variety of industries.[266] While in some cases lock-in can even lead to "close user–producer relations and joint learning and innovation for product development" (Greenstein 1997; p. 251), users in such cases will generally be dependent on the supplier, who, at least in the short run, can control such innovation, due to his superior bargaining power, especially if the user has invested into relation-specific assets, such as customized machinery.

The effect of dominant-designs and high switching-costs can be seen by the wide acceptance of the QWERTY keyboard which "made further innovations in keyboard design off-limits" (Utterback 1994; p. 32). Although in 1936 August Dvorak developed a more sophisticated, ergonomically more efficient keyboard, that balanced the frequently used letters between the two hands and loaded the strong fingers, the high costs of switching from the QWERTY standard effectively kept this "better system" of the market until today (Utterback 1994; p. 6).

The story of the refrigerator is similar. In 1930 there were two competing designs of which according to Rogers (2003; pp. 147–148) the gas refrigerator was the superior one with "overwhelming advantages". Nevertheless as the big electrical manufacturers such as General Electric and Westinghouse had invested "extensive amounts in the electric refrigerator" the inferior design prevailed: "the technology selected was one that offered the greatest profits to the corporate manufacturers, not the technology that had the most advantages for consumers."

The assertion of market power can increase switching costs.[267] Microsoft, due to its control over the PC operating platform,[268] for instance created an "applications barrier to entry" that made it prohibitively expensive for a competing operating system, such as Linux, to attract enough developers and consumers to become a viable alternative (Microsoft decision 1999; p. 15). For example, it erected incompatibility barriers between Linux and its server OS that allowed Windows NT applications to print and share files much faster than Linux files.[269] The more Microsoft files a PC user has, the more it will cost to switch away from Windows.[270] A 2003 Gartner Dataquest study therefore found that these high switching costs made changing to Linux profitable only for a very narrow spectrum of people.[271]

Microsoft has further used its "enormous [market] power" (Gawer and Cusumano 2002; p. 141) to block innovative programs by including incompatible default standards in its operating system. In a preliminary version of XP it blocked a Kodak application for modifying and editing digital photographic technology and included a 9-step process for users to switch to a non-Microsoft application.[272] Microsoft even pressurized IBM to stop promoting OS/2, an alternative operating system. If IBM wanted to continue its contract with Microsoft, it had to pay a royalty on every computer processor it sold, including those, which contained OS/2, rather than windows.[273] Judge Thomas Penfield Jackson even explained in Microsoft's Antitrust case, that "Microsoft threatened to terminate the Windows license of any OEM [Original Equipment Manufacturer] that removed Microsoft's chosen icons and program entries from the Windows desktop or the 'Start' menu [and ] . . . who added programs that promoted third-party software to the Windows 'boot' sequence. These inhibitions . . . stymied innovation" (Microsoft decision 1999; p. 99). In other words, according to Judge Jackson various "innovations [including user-innovations] that would truly benefit consumers never occur[ed] for the sole reason that they do not coincide with Microsoft's self-interest"[274] (Microsoft decision 1999; p. 207).

Another example of Microsoft quashing innovation was its reaction to the innovative Netscape Navigator in 1995. As Netscape did not want to license its browser to it, Microsoft leveraged its market power into the browser market,[275] by refusing to offer a version of Windows 98 in which its internet browser is either absent or removable. This "had the purpose and effect of quashing innovation" (Microsoft decision 1999; pp. 94–95) and destroying Netscape's business model.[276] Although Microsoft was punished afterwards[277] this will neither bring back Netscape, nor revive the lost innovation in this field.

OS/2, Lotus 1-2-3 and WordPerfect are other innovative products that vanished due to Microsoft's market power.[278] More recently, Microsoft leveraged its market power into web-based services. By tying its services to its programs, it requires users who want to use a particular service, such as instant messaging and online photo printing, to use a particular Microsoft program.[279] An internal Microsoft executive memo indicated that it intended to drive rival Real Networks out of business.[280] While we acknowledge that Windows also enabled a lot of third-party innovation, such as ironically the Napster client software, innovation was only allowed by Microsoft in areas where it was not attempting to obtain a market dominating position.[281]

Various other companies have also used their market power to burden third-party innovation.[282] Utterback (1994) explains that established industry players often, due to their huge sunk costs in the current technology, are reluctant to develop or adapt radical innovations. Take the example of hydrogen fuel-cell vehicles, which could eliminate tailpipe emissions altogether.[283] In 2001, The Economist already wrote that

> fuel cells are feasible today. The only reason they haven't been much used in cars is that established car maker have invested [hundreds of billions of dollars][284] in conventional manufacturing technology. This has made them reluctant to embrace radical approaches. (The Economist 2001b)

Ford executives stated that in 2003 internal hydrogen combustion engines "now have the reliability and long-term durability of gasoline engine[s]" (Professional Engineering 2003). Ballard Power, the world leader in fuel cell technology, already produced a 10-meter long vehicle with 125 P.S. in 1993. Two years later, a 12-meter bus with 275 P.S. could run for 400 kilometres before needing to refuel.[285] US Energy Secretary Spencer Abraham concurred by explaining that letting the US car- and oil-industry develop hydrogen technology "on their own would take too long" as "obviously there are some who are very comfortable with the status quo" (Ernsberger 2003; p. 5). Given the record profits of recent years—market leader Exxon-Mobil earned over $40.6 billion in 2007, the highest annual corporate profit ever[286]—the lack of an incentive to embrace competing technologies is understandable.[287] The fact that the US economy today is slightly more dependent on oil than it was in 1973, the height of the oil crises,[288] while other fuels, including renewable ones, having since been developed, indicates how successful the oil industry, through its political and economic power, has been in keeping the world reliant on oil.[289]

The car industry was similarly reluctant to adopt new engines. Only after being forced by governmental regulation, did they implement, previously existing technical improvements that enabled lower emissions.[290] In this manner they could prolong outdated technology and limit the development of new innovations. Historically, the car industry has often slowed down innovation as exemplified by "its unseemly resistance" to seat belts, air bags, emission-control devices or advanced tires (Ness 2005; p. 4).[291]

Chandy and Tellis (1998) found a similar reluctance of established companies to embrace radical innovation, despite this being sensible and even necessary for long-term success due to the fear of "cannibalizing" upon their own prior investments. They found that the higher such "specialized investments," the higher their resistance. Harrigan (1985) adds that 'exit barriers' often prevent firms from repositioning themselves to serve innovative customers, as they are unable to retrieve the value of their high investments when they leave their industry.

To summarize, standards, designs and market power can be strategically used to control user-innovation as to some extent they allow "innovation gatekeeping—controlling whether or not an innovation is diffused to an audience of potential adopters" (Rogers 2003; p. 156).

## Organizational Barriers

A further potential market barrier to user-innovation is the difficulty incurred in organizing such innovative activities. As innovation involves not just the improvement of a product, but also its diffusion, a basic degree of organizational and marketing skills will generally be required. The best illustration of this is open-source software. Despite of its alleged benefits in testing and improving software[292] various commentators have argued that the open-source movement finds it difficult to solve fundamental architectural problems and to start real innovative projects.[293] Open-source authority Eric Raymond (1999) even states that "one can test, debug, and improve in [Open-source] style, but it would be very hard to originate a project in [Open-source] mode." Many open-source programs are therefore merely further developments of previously existing software: Linux of Minix[294] or Firefox of Netscape's Internet Browser. The Economist (2006a) even proposed that most open-source projects fail to result in a marketable product. In the same article, Steven Weber, who authored *The Success of Open-Source* (2004), commented that "Linux is good at doing what other things already have done, but more cheaply - but can it do anything new?" The Economist (2004a; p. 19) added that "in software, all that has been developed are functional equivalents of proprietary software—operating systems, databases, and so on—that are sometimes slightly better and sometimes glaringly worse than their proprietary counterparts."[295] The Economist (2006a) furthermore added that of the roughly 130,000 projects of the world's largest open-source software development community, SourceForge.net, allegedly only a few hundred are still active, of which only a few will ever lead to a useful product.[296] Another noteworthy point is that the know-how required to develop open-source solutions is considerable. As many IT operators can install system but are not trained in programming, many companies have had to acknowledge that their previous belief in being able to complete open-source projects independently was overly optimistic and are now gradually reverting back to employing external parties for this purpose.

Dahlander and McKelvey (2005) in their study of 110 skilled computer engineers found that few users were actually able to make useful improvements: Only 5 responded to have improved the Linux Kernel and 9 to have enhanced a Linux application. Of these only 5 improvements were finally diffused, indicating that the remaining modifications were not of sufficient value to be passed on to the rest of the community.

While Voss (1985) in his study of applications software innovations in the UK, found that 32% of innovations in this realm came from the "user active" category, of these a third were actually developed by the supplier (12%) and further ones diffused and commercialized by suppliers (7%). Therefore only 13% were pure user-innovations, despite of the relative simplicity of this type of software at that time. While the technical abilities to innovate were often available, organizational difficulties, such as inadequacies in transferring know-how between different departments where the expertise resided, led to such a low rate of pure user-innovation.

Thirty-five percent of respondents to a German ZEW study that analysed the barriers to innovation in West Germany between 1980–2002 considered that the high effort needed to innovate was a critical barrier.[297] The lack of capital and problems in turning technical know-how into products were also identified as sizeable barriers. As to financial restraints Åstebro (1998) determined that independent inventors (in this case a proxy for user-innovators although these are not necessarily the same) are only about 17–25% as likely to commercialize their inventions as inventors in established companies. Finkelstein et al. (1995) commented on the difficulty for physicians to finance their experiments as investors were often reluctant to support these. Leifer et al. (2000; p. 110) in regard to radical innovation determined that in 75% of the projects they evaluated "external financing made the difference between project continuation and being put on the shelf." Apart from the financial costs of user-innovation, the time that could be devoted to other endeavors is also a critical limiting factor. In the Dahlander and McKelvey (2005; p. 630) Linux investigation the main reason given for not improving the code was "lack of time."

The access to the requisite complementary assets needed to successfully develop and commercialize a technology, such as adequate production and testing facilities, adequate supply arrangements or marketing and distribution channels, will be a further problem for many users. Hayes and Wheelwright (1984; p. 381) explain that the success of world class German and Japanese manufacturing companies "is the result of the proprietary production processes that are incubated in these shops and therefore unavailable to their competitors [and users]." Eize De Vries, a wind power consultant, stated, in regard to the Danish wind-turbine industry, that user-innovation involving "the use of scrap or readily available materials often results in [technically] sub-optimum solutions, unfit for series production" (Douthwaite 2002; p. 73). To illustrate take the Danish carpenter, Christian Riisage, who in the late 1970s built his own wind turbine out of old car components. Various accidents and attempts later he had constructed a 22 kW machine that he attempted to commercialize. Due to the inferior materials used and the unprofessional engineering the product proved unreliable. The resulting damage claims by twelve owners drove him out of business. Shah and Tripsas (2004) confirmed that users often struggle to commercialize their developments.

Lee et al. (2003; pp. 22–23) took these arguments a step further and persuasively argued that an "open-source system will never exist alone, though a closed-source system can." They argued that the existence of an open-source movement depended on two vital factors: a) "a critical number of high-end programs relative to its industry's innovative potential" and b) "a well-developed mechanism to transfer the signals to the surrounding environs" (i.e. the job markets must acknowledge the information discovery service of the OSS).[298]

The lack of these factors in and the resulting non-existence of an open-source movement in most other areas except software indicate that the cost and complexity of innovation[299] and the difficulty of such coordinated group action, often restrains group-user-innovation. Despite of Bruns (2001) vision of open-source nanotechnology, the medium term future of this realm, as well as of the biotechnology industry,[300] will, due to the need for scale, huge investments, increasing pace of innovation and long time horizons for payback, be shaped by "established industry giants" (Veverka 2003).

To summarize we should point to the analysis of Biemans (1991). While agreeing that users play an important role in the innovation cycle, he argues that the numerous organizational barriers that users have to overcome before being able to innovate by themselves should be considered as well: "Irrespective of the industry concerned, von Hippel's reasoning should be expected to severely underestimate the necessary involvement of a manufacturer in the initial stages of the process of product development" (Biemans 1991; p. 167). This is as manufacturers, even in cases where users approach them with a prototype, will nearly always undertake significant steps in the early-stages of product development to assure that the product meets the industries' requirements: Manufacturers typically "need to modify the user-developed original considerably" (Biemans 1991; p.167). In his study of Dutch medical devices innovations, he found that while 53% of the product development processes were initiated by a user or a third party, as soon as it came to the concept development stage, manufacturers started to dominate the process. Across all seven stages of innovation users were found to be involved in around 50% of the activities, while manufacturers participated in nearly 90%. While this result is in part due to Biemans' methodology it does support the underlying assertion that the democratization argument neglects to consider not only the influence of manufacturers but also of third parties. Biemans (1991) found that in the field of medical devices, an area where user-innovation is considered to be very high, third parties such as distributors, suppliers, governmental agencies, scientific foundations or research institutes played an equally if not more important role than users. Indeed, while users were influential in the early phases, third parties tended to be supportive throughout the innovative process. In other words, manufacturers or third parties had to fill the organizational deficits that users may have in successfully commercializing a product.

## TECHNOLOGICAL BARRIERS TO USER-INNOVATION

### The Impact of Technology

We have now seen that legal and economic barriers exist that can restrain innovative users. In a world where technology is increasingly shaping our lives, the overall influence of the previous factors will however largely depend on the constantly changing technological architecture that both companies and users face. We are progressively moving into new technological realms where new rules make the old ones obsolete. In telecommunications we moved from the electro-mechanical to the digital-optical paradigm and in pharmaceuticals from a chemistry—to a genomics-based drug development.[301] In this section we will attempt to explain the impact that some technological changes are having on user-innovation.

The democratization argument is based principally on changes in technology having facilitated user-innovation: "The steady improvements in computer software and hardware are making it possible to develop increasingly capable and steadily cheaper tools for innovation that require less and less skill and training to use" (von Hippel 2005a; p. 121). Professors Franke and Harhoff similarly argue that "new technologies wipe away the barriers between [manufacturing] corporations and users" (own translation Schreier 2005; p. V). While this may be the case in the realm of open-source software and information products, it is less certain whether such analysis is readily applicable to physical goods. While information technologies have revolutionized the creative medias,[302] or have given users the ability to design their own shoes or watches,[303] our focus is not on the creative arts but on the development of new technologies. We are therefore less concerned with design, as in the mass-customization cases, but with innovation as we defined it. In this part we will attempt to demonstrate that new technologies can, even with respect to information goods, actually increase the borders between manufacturers and users.

While previous sections largely dealt with the right to innovate, this one concerns the ability to do so. To be more precise, the two preceding sections dealt mainly with barriers that prevent users from commercializing their developments, not necessarily from undertaking these in the first place, while this section focuses on the difficulties and limitations imposed on tinkering.

At this point it has to be re-emphasized that "innovation requires the synthesis of user need and the technical means to meet the need" (Voss 1985; p. 121). Lettl et al. (2006; pp. 262–263) explain that this technical ability includes a "high competence in their domain"—i.e. a high ability to understand one's own needs and one's technological environment— "access to technological know-how," resources for the user's own R&D, the requisite tools (manufacturing instruments, software programs, laboratory equipment etc.) to implement one's ideas, and finally "technological expertise." In this respect we need to point out that just like industrial firms, users "are constrained in their search by their existing range of knowledge and skills" (Pavitt 1984; p. 353). While currently possessing the technical expertise may not always be essential, the cost and difficulty of acquisition always have to be considered and may serve as a constraint.

This section is split into two parts. In the first, we will argue that due to the constantly increasing complexity of innovation, it has become difficult for users to obtain the technical ability to innovate. In the second, we will explain that modern technology, such as software, trusted systems, genetic engineering, or nanotechnology can be used to (purposefully) impose significant barriers on users wanting to innovate.

## Increasing Technological Complexity

Improving a technology clearly requires substantially more technological competence than mere use. Benjamin Jones (2005; p. 2) therefore explained that innovators "must initially undertake significant education" to bring themselves to the "frontier of knowledge," from where they could potentially innovate. Lüthje et al. (2005) state that the lack of a user's personal technical knowledge often explains why various users have to stop their efforts at the idea/concept stage. Lotz (1992; p. 37) adds that technologies such as electronics "have very high levels of threshold regarding the knowledge required to master them." In this part we will argue that many factors indicate that it has become increasingly difficult in many areas for users to fully understand the technologies in question and to attain the skills and tools to improve it.

It is accepted that free user-innovation requires an "information commons" (von Hippel and von Krogh 2006), such as prior open-source projects in the realm of software, to allow users to freely use a wellspring of knowledge for their innovative activities. Due to the involved costs of obtaining access to external information[304] users in general will rely mainly on local information. In Lüthje's (2005) mountain biking study less than 16% of users indicated that they had "acquired new knowledge to develop the solution to their problem." We believe that the information commons will often be insufficient to provide users with both the needed tools and know-how. While the open-source software or the nascent open-source biotechnology movements are trying to increase this commons, the increased commercialization of academia,[305] the rise in the costs of R&D,[306] and the increased assertion of IPRs do not point towards a widening of the commons. Nelson (2004) argued that the upsurge in proprietary science and university patenting means that fewer results are publicly disclosed. He warns that this may destroy the scientific commons on which scientific progress is so dependent. Given the increased financial pressures on universities and other publicly funded research institutions, it does not seem that this trend will be reversed soon. We agree with Pavitt's (1984; p. 348) succinct summary: "The assumption that most technological knowledge is or could be publicly available and generally applicable has little foundation in reality."

Our first argument in this sub-section, is thus that innovation has become more complex. Many renowned commentators have emphasized this point. Schiffer (2003) for example, explains that while during the mid to late 1700s hobbyists were able to teach themselves a science in usually just a few months, this was no longer the case by the 1800s. Consequently, while the founders of electrical science, such as Benjamin Franklin, were nearly all amateurs, inventors of the late 1800s, such as Thomas Edison or Sir William Siemens, were already largely professionals.[307] Kingston (2004) similarly suggests that since the 19th century invention has primarily become the result of investment in R&D in corporate research laboratories.[308] As scientists began to require access to an ever-increasing amount of technical know-how[309] and to specialized, complex and expensive apparatus, the size and sophistications of laboratories began to grow. World-renowned economist John Kenneth Galbraith (1952; p. 86) also stated that "technical development has long since become the preserve of the scientist and the engineer."

Noam Chomsky, the linguist and social critic, further argued that modern science has stretched the cognitive capacity of humans to the breaking point, so that while in the 19th century, any well-educated person could grasp contemporary physics, today only "freaks" comprehend all that is needed to invent and innovate in this area (Horgan 1998; p. 153). Another commentator argued that scientific progress is becoming increasingly difficult and expensive: "We must push into regions never investigated before, regions of higher density, lower temperature, or higher energy. In all these cases we are pushing fundamental limits, and that requires ever more elaborate and expensive apparatuses" (Rescher 1978; p. 207). Jewkes et al. (1969; p. 179) add that while through the 20th century there were numerous sources of invention "the process of discovery and invention is becoming progressively institutionalized." While this trend cannot be generalized, we have seen large drops in the relative importance of individual inventors in certain industries such as pharmaceuticals, aeronautics, chemicals or as we shall see below, in seed-breeding.

To further illustrate the increased complexity of innovation, Allison and Lemley (2001) and Lemley (2000) use data from the US patent system to demonstrate that there has been a substantial increase in prior art citations of all patents from the 1970s to the 1990s. While in their 1970s sample the average number of references cited per patent was 5.72 in the 1990s sample this had risen to 15.16, an increase of 165%. Table 2 illustrates that there was a substantial rise in patent citations across all industries, with most increases considerably exceeding 100%.

Jones (2005) also showed that innovation has become more difficult and expensive: While both the number of US R&D workers and the real US R&D expenditures have sharply increased over the last 30 years, the US Private Sector Multifactor Productivity growth has been practically flat. The average number of patents produced per R&D worker or R&D dollar has also been falling over time across US manufacturing industries.[310] Furthermore, consider that the majority—Contractor (1985) found 75%—of licensing agreements contain know-how transfers, which users will frequently lack in their tinkering.

This gives an indication of the difficulty of understanding and using new technologies, not to mention the increased know-how needed to innovate. While manufacturers can often acquire external know-how, the high-costs and difficulties of transferring tacit information[311] can make it difficult for users to obtain the required know-how. It is crucial to point out that obtaining access to the needed know-how will be much more difficult for individual users than for corporate users who may be world-class in their particular niche area. In respect of mountain biking, Lüthje et al. (2005; p. 13) stated that "individual user-innovators in mountain biking presumably do not have major resources to devote to the acquisition of non-local solution information." Indeed, Voss (1985) found that 64% of users lacked the technical expertise needed to innovate in the field of application software. Ogawa's (1998) findings that innovations that required a

*Table 2* Growth in Complexity on the Basis of Patent Citations

| | *Medical Devices* | *Pharmaceuticals* | *Elec-tronics* | *Chemistry* | *Mecha-nics* | *Accoustics* | *Optics* | *Energy* | *Automobile* | *Communications* |
|---|---|---|---|---|---|---|---|---|---|---|
| Sample 1976–78 | 5.76 | 2.24 | 5.16 | 4.37 | 5.47 | 4.32 | 4.17 | 5.52 | 5.64 | 4.52 |
| Sample 1996–98 | 19.44 | 6.06 | 9.35 | 9.94 | 12.7 | 8.95 | 10.16 | 14.63 | 12.36 | 10.56 |
| % Δ | 238% | 171% | 81% | 127% | 132% | 107% | 144% | 165% | 119% | 134% |

Source: Allison and Lemley (2001)

high understanding of technology were often carried out by manufacturers, while users were responsible for innovations that required a high need understanding, also points in this direction.

To illustrate the effect that this increasing complexity of new technologies and product has on user-innovation, we need to consider some more examples. Several of the user-innovation studies have involved only moderately sophisticated technologies such as mountain bikes,[312] pipe hanger hardware—a relatively simple low-tech product as the authors acknowledge[313]—extreme sports goods,[314] Apache OS servers[315] or library information systems.[316]

Take the development of tennis rackets. These used to be simple wooden constructs with tight strings. In the early days lead-users frequently experimented with new strings, new racket shapes or different materials. Indeed Dahlin et al. (2004) report that the development of the tennis racket industry was largely shaped by independent inventors. Today, this has changed considerably as technological developments have made rackets very complex. First came durable metal and fiber-reinforced rackets, then powerful oversize and wide-body models, followed by finely balanced glass-fiber or graphite rackets with shock and vibration-dampening.[317] Today, tennis rackets are high-tech gadgets, developed in high-tech laboratories, consisting of high-tech materials, such as titanium or hyper-carbon—a material used in rocket technologies—using advanced manufacturing techniques. They are ultra light, super durable and have huge sweet spots, i.e. the optimal hitting area. Recently, Wilson commercialized its nCode rackets which are constructed using nanotechnology to further strengthen the frame.[318] To design the perfect racket, the material used, the weight and the stiffness of the racket frame have to be carefully apportioned, to optimize the complex aerodynamics and mechanics of the ball/string/frame collision.[319] Improving tennis rackets today requires both a substantial amount of know-how and the adequate tools (not to mention ways of getting around the patent portfolios that key players like Wilson have built).[320] Consequently, even lead-users, i.e. those who have a strong personal need for an optimal racket and would substantially benefit from an improved racket, such as professional tennis players, are unlikely to be able to improve their rackets themselves. As manufacturers are needed to control the technological feasibility, users generally only give feedback to allow their rackets to be tailor-made for their needs.

The example of lead-user Roland Sommer, a German aeronautical engineer, and tennis enthusiast, illustrates this. In the early 1990s, Sommer developed a new system for stabilizing the racket head.[321] In order to become an innovation his development needed considerable further fine-tuning. Only together with the aid of ProKennex, a large Taiwanese racket producer, and further $3 million dollars in 2 more years of R&D, was the Kinetic racket frame optimized.[322] Given that racket technology has further improved dramatically since the Kinetic's development, this case shows that user-innovation depends on access to the right tools and resources, which in most sports few users will possess themselves. What professional soccer-player will be able to improve the current finely balanced footballs, such as Adidas' TeamGeist?[323] Which professional golfer or which professional Formula 1 driver will be able to improve his clubs or vehicle himself?

Hienerth (2006) found support for this assertion in the rodeo kayak industry. While the industry was created by users who actively exchanged their designs and new creations, as the industry matured and users started to found commercial enterprises such free exchanges became increasingly infrequent until they only occurred in regard to non-strategic issues. Furthermore, once the technology matured, in other words once the shift from fiberglass, which users could inexpensively and easily modify, was made to plastic rodeo kayaks, user-innovation substantially declined. It simply became too technically demanding and costly to improve such kayaks. Therefore in this user created industry since 1993 no homemade kayak has entered into a world championship competition.[324]

These examples illustrate that in most sports individual users will be unable to provide anything else except general feedback on use. Innovation will occur by groups of sophisticated engineers who possess the adequate training and tools. Rather than a democratisation, in most long-established sports the separation between manufacturers and users has become clearly established.

Innovation has also become more arduous for toner cartridges users, as embedded chip technology has become increasingly complex. While in the late 1990s "a single technician with minimal equipment [and resources] could conduct the entire [reverse engineering and adjusting] operation," today many months are needed in which highly trained "team[s] of specialists from a half-dozen disciplines, combined with automated, robotic production machinery" are needed to get around the chip technology to innovate in regard to ink cartridges (Golden 2002; p. 40). For instance, developing an appropriate aftermarket solution for the HP 4100 Printer took nearly 13 months.

The prohibitive costs of obtaining access to the needed tools can be seen on the basis of rapid prototyping tools. These technologies driven by CAD data, which produce physical models and parts, are supposed to decrease the costs and the time taken for developing prototypes. Von Hippel (2005; p. 123) drives forward his argument by stating that "the solution components now available to users are often as good as those available to professional developers." With a minimum cost of tens of thousands of dollars for low-cost 3D printing devices[325] such an observation, at least for the time being, seems to be limited to information goods and the creative arts. As technological complexity progresses such low-cost 3D devices will generally be insufficient and more complex systems may be required such as "enterprise prototyping centres" that are more suited for developing viable prototypes and these typically cost 2-25 times that of a 3D printer.[326] Such machines will only be a suitable investment if they are used frequently, which is unlikely to be the case for all users who do not happen to be professional developers.

The realm of nanotechnology further illustrates that both the expense of needed tools and the difficult access to them can amount to a substantial hurdle for users. While Bruns (2001; pp. 203–206) believes that in the long run open-source business methods could be applied to the development of nanotechnology innovations, currently the essential tools, like atomic force microscopes, are "still expensive equipment sold individually to scientific laboratories and industry. The cost of hardware [is thus] . . . well out of the hobbyist range." He acknowledges that so far, due to both time and complexity concerns, all nanotech open-source initiatives have failed to take off.

A quick look into the biotechnology industry gives us further insights. Jennifer Hope in her doctoral thesis *Open-Source Biotechnology* describes how open-source concepts could be implemented in this realm. While she comes to the conclusion that open-source biotechnology is "both desirable and broadly feasible" she does accept that currently a proprietary culture still dominates the field: While "the norm used to be one of open-collective invention, this culture has been largely replaced by complex proprietary networks" between the public and private sectors (Hope 2004; pp. 227 and 179). Today, "pharma live, breath and die on patent positions. . .not sharing IP is deeply ingrained" (Lee Bendekgey, Senior Vice President and General Counsel of pharmaceutical company Nuvelo, cited in Hope 2004; p. 170). The differences between the software and biotechnology industries are that in the latter development is more expensive[327] and difficult, the R&D cycles and the pay-off periods much longer,[328] diffusion requires more than a fast internet connection and most critically "fewer people have the necessary level of skills and commitment to conduct biotechnology research and development. . . [and] fewer people have access to the necessary infrastructure and supplies" (Hope 2004; pp. 164, 166 and 188). In order to innovate in biotechnology, parties require access to "highly sophisticated and expensive equipment," and to master the "increasing complexity of knowledge and research [in this realm]" they require expertise in "pharmacology, medicine, genetics, chemistry, information technology and biology" (Suarez-Villa 2004; p. 455). As Roger Brent, Director and President of the Molecular Sciences Institute at the University of California, San Francisco, summarized "it's harder to build things in biology than it is to write code. It's a great deal harder, and it's slower. You really can't do it part-time; the skill level at this point is high. For fluency in nucleic acid manipulations, I'd say the typical person here has had eight years of post-undergraduate education" (Hope 2004; p. 190).

The disillusioning realization that only a few larger biotech companies have ever managed to be profitable and that the majority of small or medium-sized biotech companies "barely end up covering their expenses and often manage to survive only by establishing collaborative links with other firms so that costs can be spread about" (Suarez-Villa 2004; p. 456) furthers our doubts about open-source biotechnology. Although Maurer et al (2004) propose an open-source approach[329] to develop cures for tropical diseases, they acknowledge that "nothing will happen unless Western governments and charities foot the bill" (Maurer et al. 2004; p. 36).

There are also some indications that technological complexity has affected the software industry. Innovation in this area is generally cumulative, meaning that access to prior software programs is often a prerequisite. Creating new programs from scratch is often too expensive and time-intensive. In order for users to improve software they require access to source code. While originally the software industry started with the free and open distribution of source-code, especially due to the involvement of academic and governmental research laboratories, this changed in the subsequent years as an increasing amount of companies enforced their IP, and started to impose copy-protection mechanisms.[330] Although an increasing amount of open-source software is available, closed proprietary software is still the norm.[331] Stallman (2002) stated that around 90% of users don't yet use free software, which is subject to the GNU General Public License (GPL) and thus copylefted so that both its current and future use remains free.[332] As large software companies want to sustain and enhance the "rigid divisions between product design and product testing" (Neff and Stark 2004; p. 178), the source code to many key software programs, such as Microsoft's, is still largely closed. To modify a closed code software users first have to reverse engineer the code.[333] As Johnson-Laird (1994) and Samuelson and Scotchmer (2002) explain reverse-engineering of closed source code is so difficult, expensive and time-consuming that it is rarely efficient when attempting to build a competing but non-identical program. Dahlander (2006; p. 20) adds that "software is also protected by being distributed in compiled form that makes it almost impossible to interpret and reverse-engineer."

The evolution of the complexity of many programs reinforces this. Jon Udell (2003; p. 48) explained that "the arc of software progress is defined not by increasing speed or capacity, but by the growing complexity of the data, events, messages, and relationships flowing through software systems." The lines of source code of Microsoft's operating systems have reportedly increased from 6 million for Windows NT 3.1 to 40 million for XP and over 50 million for Vista.[334] Indeed, computational models in practically all areas from financial prediction instruments to biological systems, have become increasingly sophisticated. Such resulting complexity has, according to Johnson et al. (2004; p. 65), at least in the area of biomedical research, "become a major hindrance to [the] further use of advanced computing tools," that, given both the time involved in and the need to possess the necessary computing skills, has made it very difficult for many researchers to adapt and modify such software themselves. A main contributing problem is that even small modifications can require the whole software to be rewritten. Furthermore, without access to the source code commentary, which users rarely have, it is often difficult to understand the source code adequately.[335]

To add to this, even open-source software is rarely completely open· More generally, the free-revealing, i.e. the voluntarily giving up all IPRs to some information and granting free access to all interested parties,[336] that some commentators so enthusiastically discuss in various industries[337] is generally not mainstream commercial behavior. Rather selective revealing seems to characterize corporations working with open-source software. Henkel (2004) in his study of free revealing of open-source software in the embedded Linux industry found that on average firms reveal about half (49.3%) of the code they develop. Know-how that constitutes a competitive advantage, as opposed to generic code, will not generally be made freely available. Henkel (2004) found that the vast amount of firms working with embedded open-source software (OSS) reveal generic code and protect competitively important IP. This trend is exacerbated by the fact that an increasing amount of products contain software.[338] Modifying such embedded systems, i.e. hardware-containing software, will often require more than merely computing skills. This trend that cross-disciplinary, multi-invention products are rapidly becoming the norm in high technology industries is likely to further reshape the innovation landscape.[339]

Technological convergence, where nanotechnology, biotechnology, information technology and cognition are moving closer to one another, is indeed "currently progressing at a rapid rate" (National Science Foundation 2002; p. 1). The EC Report *Converging Technologies* indicates that this convergence "promises to transform every aspect of life" (Nordmann 2004; p. 10). The Electronics, Information and Telecommunications sectors or the Agrochemicals, Pharmaceuticals and Biotechnology industries are for example becoming increasingly intertwined. The practical consequence of these developments is that users wanting to innovate would need to master a wide variety of different know-how. Jones (2005) offers some empirical manifestations of this: The educational burden on innovators, i.e. the know-how they need to acquire, has increased over the last 20–30 years. This has resulted in a considerable increase in the age at which Nobel Prize winning research and great technological inventions are made (an increase by 6 and 5 years respectively over the 20$^{th}$ century) and a large drop in the amount of great inventions that are developed early in a person's life cycle.

For users, it specifically means that in order to be able to innovate, researchers, doctors or farmers would need to master not only their own fields but also other skills such as software programming, technical- or genetic-engineering. Douthwaite (2002) explains that many if not most mature products such as the current generations of wind turbines, motor cars, nuclear power or pharmaceuticals have become too complex for most users to adapt and modify themselves. While such complexity can be overcome by the establishment of multidisciplinary teams of complementary specialists[340] this requires professional dedication. The substantial increase in the number of average inventors per patent—since 1975 there has been a "remarkably steep rate of 17% per decade" (Jones 2005; p. 4) from around 1.7 to close to 2.3[341]—illustrates that there has indeed been an increase in group innovation.

There has also been a sharp increase in the assignment of patents from inventors to corporations. While at the beginning of the 20th century around 18% of patents were granted to corporations, this increased to 79% by the mid 1970s[342] and to 85.1% by the mid 1990s.[343] Allison and Lemley (2000; p. 79) therefore stated that in the US "patents tend to be granted to corporations and to collaborative groups of inventors, not to individuals working alone."[344] Dahlin et al. (2004; p. 2) referred to this as the "industrialization of invention."

At this stage, we need to consider a counter effect of complexity, namely that it may actually increase user-innovation. This may be as in many industries R&D has simply become too complex for one firm alone.[345] The resulting focus on core competences[346] and increased reliance on networking and contracting with third parties to obtain access to innovation in the software, chemicals, biotechnology and semiconductor industries, may therefore offer new opportunities for users.[347] In the open-source software movement for instance thousands of programrs with the widest possible degree of complementary skills, cooperate to overcome the technological difficulties that burden individual user-innovation.[348] If companies design their products in modular ways for example, this could give users the "freedom to experiment with product design" (Baldwin and Clarke 2004; p. 152). In regard to semiconductors Arora et al. (2002) indicate that the increasing complexity has led to the segregation between firms which produce chips and those that design chips. These design firms, which have been described as users, have been a significant source of *design* innovation in this field.[349] This actually supports our view as such design firms innovate because it has become their professional role in the industry's value chain to create new innovative chip designs which manufacturers will then create. We are thus seeing a trend of the professionalization of user-innovation. Dahlin et al. (2004) support this contention. Although they found a considerable degree of independent inventors in the tennis racket industry between 1981-1991 "hobbyists", non-habitual inventors who are not purely driven by financial rewards, constitute only a minority. The average independent inventor held 8.32 patents, indicating that these were not lay but instead professional inventors. What is more significant is that the hobbyist inventors were much more likely to hold low-impact patents than any other party.[350] Dahlin et al. (2004; p. 1180) conclude that in other industries there will be even higher "barriers to independent invention," leading to an even lower degree of independent inventors.

Kodama (1995; p. 145) supports these contentions by explaining that von Hippel's analysis is "limited to innovation in industrial goods, such as scientific instruments and semiconductor process equipment. In the area of consumer goods, it is hard to imagine end users as a source of innovation." Veryzer (1998a, 1998b) found that in regard to radical innovations users may often be less suited due to the involvement of advanced and often proprietary technologies[351] and may even resist such efforts.[352] Furthermore, if companies listen too closely to existing customers they can even fail to seize the opportunities of new technologies.[353]

Although we don't completely agree with these simplifications as even consumers can sometimes generate innovations and as sometimes users can be a source of radical innovation,[354] they highlights the important argument that involving users into a companies' innovative activities is a question of corporate strategy and that there are certain limits to user-innovation imposed by technological factors. Sony's vigorous pursuit of mod chip creators, while at the same time opening up its PlayStation game consoles to the Linux operating, illustrates that even within one company the extent of user involvement can differ substantially.[355]

To summarize, in this section we explained that innovation is becoming more complex. For users, this poses a barrier to innovation in that both an increasing amount of know-how—the rising "frontier of knowledge" (Jones 2005; p. 2)—and access to ever more sophisticated tools will be needed. Together with the progressing technological convergence this development has the potential to exacerbate the technological gap between corporate and individual users.[356] User-innovation is still possible however if many individuals collaborate with one another.[357] While through such collaboration and entrepreneurial efforts the needed know-how and tools may be obtained, in order to overcome this barrier to user-innovation, we argue that users will increasingly require professional dedication.

## Technologies to Restrict User-Innovation

While the law may be a powerful instrument to discourage users from innovating if companies so desire, it may often be a cumbersome and tedious process to stop those who have already breached a contract or infringed a patent. Harvard Law Professor Jonathan Zittrain (2000) thus argues that as companies will become increasingly dissatisfied with the expense, long durations and administrative difficulties of the legal processes, especially in regard to individual users where small amounts are at stake, they are increasingly likely to employ technological means to attempt to appropriate their R&D investments. As Rogers (2003; p. 184) phrases it, "some designers of an innovation structure a new idea so that it is particularly difficult to re-invent." Consequently, apart from having to overcome the inherent technological complexity, users wanting to innovate also have to deal with artificially erected technological barriers aimed at controlling product use.

The main difference between legal and technological means to regulate behavior is that the latter operates ex-ante, i.e. before a successful modification has occurred. To illustrate consider the example of databases. These are generally designed so that only authorized users can utilize them. The technological mechanism of keeping unauthorized users out such as digital encryption, which are becoming increasingly sophisticated,[358] operate in an "automated and self-executing" manner to distinguish between authorized and unauthorized users before the undesired access occurs (Reidenberg 1998; p. 568). The law on the other hand, in this respect suing for trespass or database interference, generally only regulates ex-post or retrospectively. Although the law may have a prospective deterrence value, this only applies to certain types of users.[359] Legal ex-ante instruments, such as preliminary injunctions, unlike technology, require both prior information of an impending interference and court action. A further disadvantage of legal rules is that they generally evolve too slowly to accommodate to rapid technological changes[360] as illustrated by the slow evolvement of computer security laws.

While a wilful infringer can in theory circumvent both forms of regulation, with technology he at least requires a skill, e.g. hacking, that many users may not possess. Although some commentators have argued that any technological barrier can be circumvented by sophisticated hackers[361] this does not mean that the average user is capable of overcoming them. On the contrary, even lead-users in their select areas are likely to often be prevented by technological measures simply because they do not possess the time, financial resources or ability to successfully circumvent such measures. This would require them to either be highly sophisticated hackers themselves, or to hire such experts.

This refers to one of the key arguments of Lawrence Lessig's *Code* (1999), where he argues that technological architecture, what Reidenberg (1998) refers to as "lex informatica," can be a much stronger way of regulating behavior and protecting IP than the law. This is especially as unlike IPRs, technology is not restricted to a particular territory, and can regulate globally. Given the increasing digitalization in most industries,[362] such technological regulation is increasingly enabling "rights holders [to] effectively write their own intellectual property statutes" (Burk 2004; p. 1565). While this is already possible in the realm of software, and will increasingly be feasible in the realm of biotechnologies with the manipulation of DNA, the argument has been made that nanotechnology will allow the programming of matter[363], increasing the possibilities of developers to assert control over their products. Technology thus has the potential to "supplement (or eventually replace) intellectual property as a means of strengthening corporate dominance over products and processes" (ETC Group 2001; p. 5).

## Software

The realm of software, a paradigm area of user-innovation, seems to be a good place to start scrutinizing how technology is being employed to restrict user-modification. One increasingly important tool to control software users is "activation." This anti-piracy tool, by extracting the ID numbers of a computer's hard drive/random access memory (ram) and matching these to a specific version of software, is intended to ensure that only authorized users can access a program.[364] In this way installation of the software is intended to be limited to just one computer. Microsoft for instance can, in theory, "block each copy of XP from being used on a second computer" (Mossberg 2001). This strategy of forcing each computer to have its own independent, and separately purchased, operating system on it, "is what made Microsoft rich" (Douthwaite 2002; p. 122). Furthermore, an activated copy of Microsoft's operating system, XP, can track which PC it is on and can shut itself down if a user modifies his hardware configuration too much.[365] Another example of activation is the Casio Cassiopeia BE-300 PDA. If a user wants to use Microsoft Windows CE on it, he has to register the software within 30 days after installation or his PDA will shut down.[366] As activation, which is inter alia already a part of Office XP, Macromedia, QuarkXpress and Adobe Photoshop,[367] can increase software companies' profits substantially,[368] it is likely to be incorporated into an increasing amount of software programs.

If activation works as intended, it has the potential to limit the use of software, at least to some extent, to authorized users and thereby decrease the population of tinkering-users. Although currently this technology cannot fully prevent pirating, Microsoft, by requiring users who update their software to certify that their software is genuine, can prevent users of pirated software from updating, thus eventually forcing these to obtain a new copy of their software.[369] A further simple means to restrict users' ability to tinker with software is simply not to release its source code. Microsoft is the best example of this. Until it was forced to disclose parts of its Windows' source code,[370] it was able to limit improvements to its operating system and its office programs by all non-authorized parties. Dahlander (2006; p. 20) thus argues that innovation in respect to proprietary closed software "takes place within the boundaries of the firm or is supplemented by contractual relations with suppliers and consultants. Customers cannot further improve the software, but merely can send suggestions to the firm with new functionalities and demand or hope that the vendors fix the bugs."

Obfuscation is another important barrier to software innovation. This refers to the creation of entropy to a program's instructions. By adding extra code and transforming existing code so that it is difficult for humans or code analysis tools to understand, creators can increase the difficulty of unauthorized modifications. Lie (2003; p. 2) however pointed out that "software obfuscation is only effective against an unsophisticated adversary", as advanced parties will be able, given enough time and resources, "to de-obfuscate the program."

Encryption is a sophisticated version of obfuscation. This process of scrambling data to make it unreadable by anyone other than the intended parties has indeed been considered to be one of the most powerful tools to control software use.[371] Encryption is used to limit access to source-code, to databases or networks to authorized users.[372] As non-authorized users wanting to improve encrypted code, first have to decipher it, requiring the requisite skills and resources, Samuelson and Scotchmer (2002) believe that companies can design code to make unauthorized innovation in this realm practically impossible.

Although encryption technology has made huge advances over the years,[373] hackers still often get around current encryptions, as the example of the DVD anti-copying mechanism CSS illustrates. Norwegian teenager, Jon Johansen, and his Masters of Reverse Engineering hacker group developed and distributed DeCSS, which enables the circumvention of CSS and thus the saving of audio- and video files without the encryption.[374]

The next generation of media encryption is however more sophisticated. Most Blu-ray discs, the new DVD standard, use AACS (Advanced Access Content System), a 128-bit Encryption Standard, while CSS only used a 40-bit- encryption,[375] an additional protection layer called BD+, which ensures that the discs only play on authorized Blu-ray players. Although AACS has already been cracked, BD+ has not yet been compromized.[376]

This is just the beginning of tamper-resistant products. The increasing use and sophistication of biometric authentications such as fingerprints, voice recognition and retina scans or the implantation of chips into the human body[377] will further increase the difficulty of unauthorized access and use of a product.[378]

Quantum physics will also play an increasingly important role in encryption. Although, quantum computation is foreseen to be fully operative only by 2020,[379] MagiQ Technologies is already selling its "Navajo" encryption machines.[380] Such a quantum machine uses individual photons to transfer the numeric keys that are used to encode and read secret documents. As these discrete hyper-sensitive energy particles react whenever anyone tries to spy on their travel from one point to another, the sender and the recipient will be alerted and the stolen code can be invalidated.[381] As the Navajo system continuously changes its key it is supposed to be nearly impossible to crack.[382]

## Hardware and Software

On a wider level, due to the increasing combination of hardware, software and encryption technology,[383] the use of an increasing amount of products can be controlled more sophistically. Many hardware developers similarly desire to maintain complete control over their systems.[384] Advanced product identification methods can for instance be employed to identify illegitimate components/copies. DuPont's Bio-Molecular Markers, synthetic DNA ink containing a code that contains a lock that has to be opened by a special decoding pen, can allegedly not be reverse engineered.[385] The DNA markers, which can provide absolute scientific proof of the identity of a product, have already been embedded into inks, paper, labels, textiles, plastics, metals, and microchips due to their potential to allow corporations to authenticate their technologies.[386] Microdots, one millimetre wide polyester particles, which can be sprayed on to components and if held under ultraviolet light, reveal the components' unique identity number are another alternative.[387] To further increase the difficulty of reverse engineering or unauthorized copying or modifying, components can be mislabelled, custom parts can be used or non-functional features can be included into the product.[388]

Modern chip technology has the potential to further control product usage and turn goods into intelligent commodities.[389] Radio Frequency Identification tags (RFIDs), which via small antennas can be used to track products, are only the beginning. Edward Felten (2006) therefore argues that "it will become economic to put secret-handshake functions into more and more products" to limit and control such products' use. The advances in miniaturization[390] further indicate that independently communicating components (hardware and software), which could make modification difficult for unauthorized parties,[391] are likely to become common. This development led Harvard economist Hal Varian (2002) to state that "new chips can keep a tight rein on consumers."

Many cell phones and PDAs for instance due to embedded authentication chips are designed to only function with the right type of battery.[392] Manufacturers argue that this is intended to ensure product safety and prevent lithium-ion batteries from exploding.[393] Customers wanting to improve such batteries, such as via hand and shoe pumps that allow you to charge your own batteries,[394] not only have to enhance such components themselves, but to also make these compatible with the rest of the product.[395]

Ink cartridges can further demonstrate this. Printing companies often sell relatively cheap printers, intending to make their money by selling expensive ink cartridges.[396] As users however were unhappy with such high costs they started to tinker with refilling cartridges, which gradually lead to the rise of the cartridge refilling industry.[397] As this user-innovation threatened the business model of the printing giants these introduced different kinds of microchips into their cartridges. If such chips detect a non-authorized cartridge the print quality may substantially decrease,[398] some functionalities may no longer work or the printer will simply seize to operate completely.[399] Such increasingly sophisticated cartridge chips, which have become common practice, make it progressively difficult for aftermarket users to reverse engineer them.[400] Golden (2002; p. 36) therefore explains that printer manufacturers are attempting to "erect barrier after barrier in an effort to severely limit or even completely eliminate the aftermarket." Although antitrust law may be relevant here, they will not apply if a company lacks the requisite market power. In *Lexmark v Static Control* a US Court of Appeal even legitimized such practices by saying that "manufacturers of interoperable devices such as computers and software, game consoles and video games, printers and toner cartridges, or automobiles and replacement parts may employ a security system to bar the use of unauthorized components." While user-innovation in this realm has not been eliminated, the aftermarket has been forced to professionalize its tinkering, as the wide variety of independent cartridge refilling companies that have been formed illustrates.[401]

Numerous other products contain chips to control usage. Many game-playing consoles, such as the Playstation 2 or the Xbox360, contain embedded chips and authentication software to ensure that the memory cartridges were made by Sony and Microsoft respectively, rather than by low-price competitors.[402] When Microsoft launched its Xbox console in 2001, hackers raced to install Linux on it, which transformed it into a low-cost, high performance media-playback system. It hardly seems surprising that the new Xbox 360 console therefore contains significantly improved security measures.[403] While hackers have actively been trying to hack the Xbox 360 to run Linux, their attempts have been far less successful.[404] SmartXX, a group of mod-chip producers even stated that "a mod chip for the Xbox 360 makes no sense, as in this manner only 25% of the Xbox could be used ... the online-functions don't permit a modded Xbox"(Computerbase 2005). Breaking the Xbox 360's dynamic encoding, which is different for every Xbox, has also been considered to be a considerable challenge.[405] Another hacker group, who took the Xbox 360 apart, reports some of the security protections the Xbox 360 employs: The motherboard contains a lot of wires that are PCB protected which makes it difficult to "spy wires and to apply a man in the middle attack" and "some chips have nine levels of metal" (Devil 2006). Additionally, "the firmware of each console can be reprogrammed by M$ [Microsoft] using Xbox LIVE. This operation can be executed without the acceptation of the user" (Devil 2006).

In the automobile industry, many car parts also contain embedded chips. Data format lockouts have for instance been used to stop drivers from obtaining certain repairs at independent repair stations. Car manufacturers want to capture more than their current share of around 20–25% of the over US $200 billion car repair aftermarket once the car warranty runs out. The rest currently belongs to independent repair shops.[406] The increased computerization of cars has made various types of repairs dependent on access to diagnostic codes, which are already contained in emission control systems, antilock brakes, climate control, airbags and other safety systems.[407] Indeed, in 2002, a membership survey by the US Automotive Association, which represents more than 15,000 independent repair shop owners, found that already 10% of cars could not be repaired due to unavailable codes and lack of adequate code scanning equipment.[408] Although attempts have been made to introduce legislation that would require car manufacturers to share diagnostic codes with car owners and independent repair shops, currently only a voluntary scheme is in place that gives independents,[409] but not users themselves, access to the information and diagnostic tools necessary to service vehicles, in return for a license fee and a confidentiality promise. Despite of this system, a 2006 AAIA Survey of 1,000 independent repair shops found that a large amount of codes and tools, especially in regard to foreign cars, were still missing.

This is not to say that users can no longer tinker with their cars anymore. A grassroots movement in California is for example trying to develop hybrids that can be plugged in to the ordinary electric power grid. In order to modify Toyota's Prius for instance, users were faced with a closed source code. While this constituted a barrier for many users, the company energycs was able to get around this by tricking the system into believing that a Toyota-installed battery had been utilized.[410] Just like in the cartridge industry, this illustrates that while user-tinkering is still possible, the effort that is requires has increased and will now frequently require professional commitment.

To illustrate the interaction between software and hardware Lie (2003) demonstrated that code can be made completely tamper-resistant. Such "XOM code," (eXecute-Only Memory), can only be executed, but not read or modified. This is possible through storage on another processor and through the code authenticating the hardware it is running on. By placing the co-processors in physically hardened packages, tampering can be prevented. Indeed "hardware components can be encapsulated to make non-destructive disassembly [and copy] almost impossible"[411] (Pooley 2001; sec. 5–25). Consequently, the limited programming interface protects programs against software-based attacks while the physical packaging defends against hardware-based attacks. Furthermore, while the traditional problem with such solution was that it came at the cost of both decreased performance and additional hardware, Lie (2003) demonstrates that the required additional hardware will have modest effect on performance.

## Trusted Computing

While Lie's XOM solution is still rarely employed, a related idea of trusted computing (TC) is becoming increasingly common. The main goal of this project, which major computing, media and electronics companies have been pursuing for various years,[412] is to enhance security in order to deal with "the explosive growth of internet commerce" (Arbaugh et. 1997; p. 70). The inadequacy of existing mechanisms to attain a secure IT infrastructure leads various commentators to argue that there is an increasing momentum for the introduction and spread of trusted computing systems, initially into PCs and subsequently into mobile devices.[413]

The general idea of Trust Computing (TC) is that through added hardware and software components, the PC is able to authenticate trusted applications. In other words, these additional components allow each component in the platform to measure the integrity of the next component and thereby to create a platform where "no code is executed unless it is either explicitly trusted or its integrity is verified prior to use" (Arbaugh et al. 1997; pp. 65–66). The goal of this is to make hackers' actions futile by binding/locking content to a particular computer.

TC, generally in some watered-down version, is spreading frequently. For example, Trusted Network Connect (TNC), a hardware-based security technology that almost everyone in the computer industry has committed to support, was presented at the 2006 RSA Security show. It has been estimated that within the next 5-7 years "[a] billion PC's will have this [TC] capability" and that it will become the "primary way to authenticate to all services" (Steven Sprague, CEO of Wave Systems, whose software is used in the RSA prototypes, cited in Dornan 2006). The presented system, intended for enterprise security, if it works as advertized, will allow any device that does not meet the specifications of the network administrator to be banned from access to the network.[414] Another example is Apple's use of a special security chip in its x86 PowerMacs that is designed to prevent people from loading its new operating system onto non-Apple machines.[415] Intel's Vpro systems which were released on the global markets in autumn 2006 will allow the running of security software in an environment isolated from the main operating system, thereby theoretically making it tamper proof.[416]

TC has been heavily criticized as commentators fear that it will adversely affect (user)-innovation. Richard Stallman (2002; p. 115) argued that "treacherous computing" will take away control of users over their computers and allow big corporations to use their proprietary programs to decide which "programs you can run, which documents or data you can access, and what programs you can pass them to." Cambridge Professor Ross Anderson (2003a; § 2, 5 and 10) argues that as many "digital objects created using TC systems remain under the control of their creators, rather than under the control of the person who owns the machine," TC systems can substantially limit users' innovation options.

Kuhlmann (2004) emphasizes that TC will impose significant barriers, technological and legal, i.e. through restrictive licensing terms, to open-source developers. The German government[417] and the European Commission[418] also noted that the introduction of TC might increase entry barriers for competing software developers, especially for open-source ones. Bechtold (2004; pp. 90–91) summarizes this potential danger:

> If, for open-source software to be run on a trusted platform, a certification has to be obtained, the software would have to be re-certified each time after it has been altered and extended by an open-source programmer. This re-certification may be costly, take time and be an overly bureaucratic procedure. As open-source programmers probably will not have the resources to finance such re-certification, they may decide not to work on the software programme at all. The idea of cumulative innovation, which lies at the heart of the open-source movement, could be thwarted by the financial hurdles created by trusted computing certification architectures.

Another reason why TCs have the potential to raise the costs of user-innovation is that they can "dramatically increase the costs of switching away" from such systems (Anderson 2003b; p. 38). This is as TC-compliant PC's could be programmed to refuse to read files that were created using pirated or otherwise unauthorized programs. The network effect, which means that switching costs rise in line with the amount of files that can only be accessed with a trusted computing platform, therefore "will work heavily against non-proprietary and open-source formats" (Lockton 2005; p. 9).

Zittrain (2000; pp. 1212–1213) summarizes our argumentation: "A user who is anything less than a talented hacker [will be] truly constrained by the [trusted] system at the behest of whoever is the source of information it might display." In this respect his vision of an age of "Trusted Privication," where technological protection can replace the legal protection of copyright, has to be taken seriously. Nevertheless, it all depends on how the trusted computing architecture will be implemented in the years to come. If open-source developers could build their own certification infrastructure that would be trusted the discussed concerns could be partly alleviated.

## Technology & the law

Harvard law professor Yochai Benkler (2002) has argued that companies increasingly create their legal rights via contract and enforce these via technology. In a review of the U.S. Copyright Office (2001; p.164) the DMCA further hinted at the joint power of technology and the law: The trend towards "legally-protected technological measures that give right holders the technological capability of imposing contractual provisions unilaterally, increases the possibility that right holders, rather than Congress, will determine the landscape of consumer privileges in the future".

In order to further control users Katz et al. (2004) propose a strategy of combining patent law and the DMCA to successfully protect copy protection measures. The three steps necessary to implement this strategy are 1) to adopt a copy protection technology with the primary purpose of preventing copying, 2) to patent this copy protection technology, and 3) to strictly license this patented copy protection technology so that no unauthorized copies can be made. If properly implemented this can allegedly, at least legally, prevent users from copying any content protected by the copy protection technology: "If the copy protection technology is included in unauthorized copies, it is an infringement of the patent. If the copy protection technology is omitted from unauthorized copies, it is a violation of the DMCA."

The effectiveness of this strategy was demonstrated in court by Macrovision,[419] which owns the patented Analog Copy Protection system (ACP). ACP causes VCRs to make distorted copies and is embedded in virtually all DVD players. 321 Studios made and sold a popular innovative DVD copying software product - that sold over 1 million copies— DVD X Copy, which enabled users to make archival DVD backup copies of lawfully purchased DVDs, including functionality to circumvent ACP. Macrovision successfully obtained a preliminary injunction prohibiting sales of DVD

X Copy as the product could only produce two possible results: either to violate the DMCA when the copy protection is removed or to infringe Macrovision's patent when the copy protection is left intact. The ruling is especially tough as 321 Studios took extensive precautions to prevent its product being used for infringing purposes, such as requiring users to register their software or disabling users who were infringing copyrights and publicly committing itself to pay $10,000 as a reward for anybody who supplied information leading to the conviction of anyone who used 321's software to market pirated DVDs.[420] Despite 321 Studios' will to fight, 5 further lawsuits and 2 federal court injunctions eventually forced it to shut down in August 2004.[421] This combination of anti-circumvention technology, claims for patent infringement and DMCA violations[422] can therefore create a substantial barrier for user-innovation.

The most profound changes in the innovation landscape are however likely to come from technological progress at the nanoscopic level. Carbon nanotubes for example are 100 times stronger than steel at one-sixth the weight, are highly conductive of electricity and have a great deal of surface area, thereby "making them ideal for mechanical, electronic and chemical applications" (Choi 2004). Apart from enhanced functionality, nanotechnology products offer the potential for more product control. An ultra flat surveillance system for credit cards, car keys or wind screens has for instance already been created.[423] Although the technology is still at a very early-stage, experiments indicate that it seems likely that individual molecules will in the forthcoming years be able to communicate with each other.[424] This could allow developers or owners to program individual units of matter like software today.[425] Due to the difficulty of reverse engineering in this realm[426] and the expense of the tools necessary to handle molecules, this development has the potential to further burden non-authorized user-innovation.

To summarize, we saw that apart from enabling user-innovation, technology can also substantially restrain such activity. We strongly agree with Lockton's (2005; pp. 23–24) assessment that such architectures of control can remove what Edward Felten termed the "freedom to tinker" and therefore move "product sovereignty from the user and into the hands of the 'experts.'" While some users will continue to develop innovative programs if they devote enough time and resources to this endeavor, these barriers have the potential to progressively decrease the population of non-authorized users willing and able to innovate.[427]

## SOCIAL BARRIERS TO USER-INNOVATION

While in the previous sections we mainly dealt with tangible or at least graspable barriers, we will now deal with a more elusive one. We have seen that tinkering is often an integral part of user-innovation and that hacking can often lead to innovative results. Hackers were indeed responsible for a large amount of innovation in the beginnings of the computer industry, developing many innovative hardware and software products.[428] Apple's Steve Wozniak for instance started his career as a hacker. File-swapping and downloading are often important means to exchange the creative results of such tinkering, allowing a wide variety of users to obtain access to know-how and tools that may in turn help them develop innovative programs. By targeting these two activities we would like to argue that antagonistic labelling and stigmatization can, although sublimely, detrimentally affect user-tinkering.

Originally, the term hacker referred "to computer enthusiasts. . . who through clever programming, pushed a system to its highest possible level of performance" (Wellen and Reiter 1998). Princeton Professor Edward Felten (2004), who advocates the importance of freedom to tinker with the technology one owns, also explained that "hacker" used to be "a term of respect, applied only to the greatest of (law-abiding) software craftsmen". Hacking was a "term of honour" as hackers were the "heroes of the computer revolution" (Levy 1984; p. 7). This understanding changed in the early 1980s, when corporations, governments and the media "began to bastardize the term to include criminal conduct" (Wellen and Reiter 1998).[429] Despite of the distinction between legitimate and illegitimate (i.e. destructive) hacking, hackers today are largely considered to be criminals.[430] As a result many young computer wizards or underground/pirate user-innovators have been stigmatized as delinquents. Society often sees young computer wizards in the same light as vandals, criminals and virus writers.[431] Lessig (2004; pp. 187–188) argues that in the US, open software developers, such as those who are developing free software for the Brazilian government, and "modern day Walt Disney's," are often branded as "pirates."

Hope (2004; p. 221) adds that this "fear, uncertainty and doubt" strategy was purposefully adopted by some software companies, most notably Microsoft, to combat the advance of open-source software. In the same manner, media corporations have targeted users of file-sharing software. Jack Valenti, President of the MPAA, is even supposed to have said that the "net is populated exclusively by thieves, pirates, organized criminals and punk kids with no respect for decent standards of civil behaviour" (Greene 2001). Although illicit file sharing, copying and downloading will continue,[432] the corporate and governmental advertisement campaigns portraying illegal downloaders as criminals are likely to at least in part adversely affect such activity.[433]

Warnings against modifications are nothing new. Early automobile manufacturers for example "adamantly opposed [the] common practice" of using their cars as a stationary power sources, as this could damage the engine or differential gear (Kline and Pinch 1994; p. 20). Companies such as Ford therefore strongly warned their users against modifying their engines. These early warnings quickly led to formal restrictions being used on automobile and kit use via warranty provisions and even pressure on dealers. Together with the introduction of new models that decreased the need to modify, such policies seem to have been quite effective in deterring such modification.[434]

The effect of these stigmatizations is very difficult to quantify. While stigmatization and the threat of punishment may not affect all hackers, especially not sophisticated hackers with criminal intentions,[435] those hackers and tinkerers who like to be seen as law-abiding, are likely to be adversely influenced: "No one in the right frame of mind wishes to be labelled a criminal" (Mulhall 1997b; pp. 295–296). The "fear of meddling with devices" that was created may in itself constitute a "significant architecture of control" (Lockton 2005; p. 15). Indeed, the generally hostile reaction of many industries to underground user communities scared many users off, especially those pursuing 'grey' activities, where the legality is at least uncertain.[436] If in the early 1990s the environment would have been as it is today, Linus Torvalds for instance may arguably not have tinkered with Minix, meaning that Linux may never have been developed.

Another social barrier that users wanting to innovate have to surmount is resistance to change. Consider the case of Siegfried Markus, a pioneer in the development of the practical gasoline automobile. By 1875 Markus had developed a working "Strassenwagen," the closest operable antecedent to the modern automobile. As the sharp explosions of the early Strassenwagen however quickly caused strong adverse public pressure to form in Vienna, the inventor's home, Markus was soon forced to discontinue his promising experiments (Greenleaf 1961). Paarlberg and Paarlberg (2000; pp. xiv–xv) argue that one of the main reasons for the 100 year time lag between the industrial and the agricultural revolution was the "natural conservatism of farm people." Social barriers are especially high in regard to well-established technologies, as the examples of Thomas Edison and Henry Ford illustrate.[437] Ford, who initiated mass-manufacturing with his Model T, subsequently actively resisted further innovations, such as the superior Model A. Even Edison, arguably the greatest inventor of all times, similarly attempted to combat subsequent superior innovations, such as most importantly the alternating current. Although this was the far better technology, Edison even persuaded the state of New York to execute people using alternative current to condemn the new technology. Both cases show that it is often natural, both for humans and for organizations, to resist change, especially if business interests are at stake.

Before moving towards the next stage of analysis, we should briefly mention that social factors can also enable user-innovation. Although users often, especially if they are consumers, constitute a large group of individually relatively powerless beings, by colluding they can exert a substantial amount of pressure on manufacturers. Such pressure can be an especially promising strategy when there are adequate substitute products. The case of Sony's Aibo illustrates this point. Aibos are robotic pets that are designed to have unique personalities to be shaped by the interaction with their owners (Labrador 2002). Many owners love these toys and enthusiastically tinker with their Aibos. One particular user reverse engineered Aibo's digitally encrypted proprietary code, thereby circumventing a copyright protection mechanism without authorisation, and taught the Aibo new tricks, such as to how to dance jazz, which he published on his website Aibopet.com.[438] Sony rewarded the user for creating "a more talented creature" (Lessig 2004; p. 154), with the threat of DMCA liability and demanded the removal of the programs.[439] While the user originally complied, after a considerable uproar in the Aibo community, Aibopet.com was allowed to go back online. Sony now allows AiboPet to distribute his experimental software, and Sony is permitted to incorporate his new developments into future versions.[440]

Another example is the user-innovator who manipulated copyrighted fonts and evaded suit due to Agfa's fear of adverse publicity.[441] Also consider the enormous resistance to GM seeds in Europe and Asia, which as we will see below, constitute a considerable burden to user-innovation. German, Philippine or Indian farmers have for example destroyed various GM fields and attempted to damage many more.[442]

In other words, social unrest could lower some of the erected barriers to user-innovation. Schlebecker (1977; p. 652) even generalizes this behavior to state that "peasant societies have found forms of coercion . . . to prevent them from using scientific and technological discoveries." Scott (2000) in his book *Weapons of the Weak: Everyday Forms of Peasant Resistance* also argues along the same lines from an anthropological perspective. A good example here is the example of the introduction of the mechanical reaper in the Philippines. In the same way as the Luddites destroyed textile machines in the early 19$^{th}$ century, local Philippine farmers systematically sabotaged theses reapers as they feared they threatened their livelihoods. When the manufacturers and owners finally allowed the locals to utilize and to adapt them, the sabotaging ended.[443] The open-source movement is another example of a resistance movement that favors free and open user-tinkering to meet their needs and requirements. Douthwaite (2002; p. 52) generalizes these observations and states that all technologies require "finishing," in other words user-adoption. In some cases, adoption may only occur if users can customize the technology to suit their own needs.

To conclude however, David Robinson (2006) points to an interesting point: a potential collective bargaining problem. As the "freedom to tinker does not matter very much to most individuals" a general consumer protest may not be a frequent occurrence. While it remains to be seen whether such collective bargaining dilemma can in some realms be overcome, we will now explore the cumulative effect of the discussed barriers on user-innovation.

# B.2 Empirical Analysis in the Field of Medical Devices

To provide empirical proof for our arguments the medical device industry that traditionally has been characterized by a high degree of user-innovation[444] will be closely scrutinized.

## INDUSTRY OVERVIEW

For our purposes "medical devices are all types of products used specifically for medical purposes, with the important exclusion of pharmaceutical products" (Lotz 1992; p. 17). The medical device industry is characterized by a high degree of innovativeness. German medical device corporations for example earn over half of their revenues from products that are less than two years old.[445] The realm is also highly interdisciplinary with the areas of microsystems-technology, mechatronics, information- and communications technology, electronics, optical technologies, material sciences, nanotechnology and cell- and biotechnologies all playing important roles.[446] Especially the latter two, and the resulting convergence between the above disciplines, are likely to increasingly shape the industry's future.

The users of medical devices, namely professionals such as physicians or nurses,[447] play an important role in this area. By far the largest users of medical devices are hospitals.[448] Especially hospitals with teaching and research programs have consistently been found to be early adopters and occasionally even developers of new technologies. The development of endoscopy and lasers[449] both resulted from the experimentations of hospital clinicians/university professors and their subsequent collaborations with manufacturers. Users entrenched in an academic hospital setting have clear advantages in accessing resources, especially to "interdisciplinary know-how" (Lettl et al. 2006; p. 259) over individual practitioners. Eleven out of the 34 innovations studied by Shaw (1985) were developed either in a university or at the Medical Research Council. In 3 out of the 4 case studies of radical user-innovation in the medical equipment industry that Lettl et al. (2006) describe, university professors played a critical role. Lüthje (2003) found that 39.2% of responding German surgeons previously had an idea for a new surgical device. While 13.7% stopped after having creating a drawing of their idea, 22% either developed a prototype or a marketable product.

Fifty-three percent of the 34 medical equipment innovations commercialized by British firms that Shaw (1998, 1985) analyzed were user-dominated. In these cases users generally had built a prototype which they subsequently developed and marketed in collaboration with a corporate partner. In other cases, such as in the example of the flexible gastroenterological endoscope, users were found to not only have developed the first prototype but also to have solved some complicated manufacturing problems and to have played a critical role in the products' commercialization, by acting as demonstrators and promoters of the equipment or as presenters of scientific papers.[450]

Users were found to be especially important in the (clinical) testing of new products. In Biemans' (1991; p. 170) study users were involved in prototype testing in 75% of the cases. The "stickiness of the complex hospital setting" and the resulting "high degree of uncertainty regarding product functions" (Lotz 1992; p. 23), are probably the main justification for this finding. The fact that the "ultimate effectiveness of a device is often crucially dependent on the skills of the practitioner using or implanting [it]" (Gelijns and Halm 1991; p. 9), is a further distinguishing characteristic of the industry. Gelijns and Rosenberg (1995; p. 8) therefore summarized that in this realm there has been "a drastic departure from what might be regarded as the 'normal' division of labour between academics and private industry."

Nevertheless, the role of manufacturers should not be underestimated. Physicians are rarely able to innovate completely by themselves. The multiple and continuous interaction of users with manufacturers is instead "the prime characteristic of the process of developing medical equipment" (Shaw 1985; p. 45). Such interaction occurs at most stages of the innovation cycle, although users often played a considerably less important role in product specification and marketing than in prototype testing and development. Lotz (1992; p. 104) further explains that "only very seldom would users challenge the producer's position as the manufacturing party . . . Hospitals almost never start production of material product[s]."

Biemans (1991; pp. 164 and 167) in his study of 17 innovations in the Dutch medical devices industry further found that while users contributed to all but one stage of innovation (users did not participate in the trial stage, where the design, the trial production and the marketing plan were finalized) "their actual contribution was frequently only limited." Third parties (distributors, suppliers, universities, governmental agencies, scientific foundations or research institutes) were found to be at least equally if not more important than users in contributing to the studied innovations. While during the early stages such as idea generation and concept development users often played a crucial role, during the final stages of the innovation cycle, the interaction of the manufacturer with third parties was found to be more important and intense than with users.

A further important finding of Biemans' (1991) study is that the identity of users that were active during the different stages varied substantially. While one user was involved in the idea generation process, others were active in product testing. This indicates the overarching organizational role which manufacturers frequently seem to undertake. This can further be seen by the overall role that manufacturers played in the innovation process as compared to users. Taking all of the 7 stages Biemans considered—Idea generation; Preliminary Assessment; Concept development; Product development; Testing; Trial; Launch—users where involved in around 50% of these to some extent, while manufacturers were involved in nearly 90%.

While users obviously play an important role in the new product development in this industry, there seems to be little evidence for the contention that users are increasingly able to innovate by themselves. Instead there seem to be various factors that inhibit users' role in generating innovations. We therefore believe that it is worth scrutinizing the specific barriers to user-innovation in this field.

## BARRIERS IN THE MEDICAL DEVICE INDUSTRIES

We will now proceed to briefly explore the barriers to user-innovation in the medical device industry.

### Patent Barriers

Unlike with pharmaceuticals, where the drugs themselves are patentable, in the realm of medical devices, only the basic principles that are embodied in the devices can be protected.[451] Patent protection for medical devices is therefore only moderately strong as various instruments can often be designed in different manners. Nevertheless, patents, especially in the form of patent fences, play an important role in this industry:[452] 70% of respondents to a survey of the German Medical Device Industry responded that patents were crucial for the strategic positioning in the industry.[453] Arora et al. (2003; p. 22) found that a "patent premium," i.e. an expected positive premium gross of patent application costs that can be obtained from patenting, existed in no other industry except medical devices.

A report by the German Ministry for Education and Research (BMBF) found that in various niche areas of medical devices, such as medical imaging or miniature laser-surgery, patent thickets had formed that potential innovators would have to navigate. Alone in the increasingly important realm of the "biocompatibility of implants" 10,800 different patents were identified globally (BMBF 2005; p. 445). Allison and Lemley's (2001) study further demonstrates the forming of patent thicket in this area. While in their 1970s sample the average number of patents referenced to per medical device patent was 5.76, in their 1990s sample this had increased to 19.44. Such 238% increase over a 20-year period is significant not just in itself, but also because it far outweighed the percentage increase in other related areas such as optics, electronics, chemistry or pharmaceuticals, which can be seen in Table 2.

Nevertheless, in regard to medical devices individuals have been able to protect their own developments much better than in other areas. Lemley (2000; p. 38) in a sample of 1,000 US patents found that, individuals patented 25% of medical device inventions. While not all of these individuals will be users, it does indicate that small parties can overcome the patent barrier better than in other industries. Arguably this is because the industry as a whole is characterized by many smaller and few larger dominating entities.

## Regulatory issues

The second legal barrier to user-innovation in this realm is the increasingly bureaucratic environment that threatens to "ossify the innovation process" (Finkelstein et al. 1995; p. 139). As the precise regulatory hurdles that have to be overcome have been extensively discussed elsewhere[454] we will merely point to some aspects that we consider relevant for our purposes.

First of all, the commercialization of any medical device, at least in the EU, the US and Japan, requires the authorisation from a respective governing body such as the FDA. The European Directive 93/42/EEC for

*Table 3* Brief Comparison of European and US Product Classifications

| *EU* | *US* |
|---|---|
| CLASS I: products with low risk, the most non-invasive products and reusable surgical instruments (e.g. stethoscopes, spatulas, . . . ) | CLASS I: Low Risk—devices that present minimal potential for harm to the user. (e.g. Elastic bandages, stethoscopes or dental burs). |
| CLASS IIa: active and non-active products with a medium risk, invasive and non-invasive products for short use. | CLASS II: Moderate Risk—must conform to applicable industry standards. (e.g. mamographic x-rays systems, hypodermic). |
| CLASS IIb: active products with medium risk, that emit substances or energy with a potential risk, and products for longer use (e.g. X-rays, contact lenses). | CLASS III: High risk—most stringent category. Generally reserved for life-supporting/sustaining devices or of substantial importance in preventing impairment of human life". (e.g. implantable pacemakers). |
| CLASS III: products with high risk or those who come into contact with the vascular system or the central nervous system (e.g. vascular implants). | |

Source: BMBF 2005, Termini 2004, Boeckmann and Frankenberger 2002, Whitmore 1997

example requires that some kind of a clinical evaluation is undertaken for every medical device that is commercialized. Not every clinical evaluation however requires clinical investigations—in some cases an active literature review may be sufficient.[455] The extent of the required formal approvals depends on the type of medical device involved[456]. While in the EU products are classified into risk classes depending on their invasiveness, in the US and Japan the potential risk to the patients is decisive[457]. As a general rule, the higher the risk class, the more difficult it will be to obtain marketing permission.[458]

Various studies and commentators have confirmed that such regulations while necessary to safeguard patients constitute a "high hurdle . . . in the innovation process" (Finkelstein et al. 1995; p. 141). A Spectaris (2004) study of the German medical device industry found that 95.2% of respondents indicated that such laws and regulations were endangering the industry's growth. The BMBF (2004; p. 521) study further confirmed that the long duration and expense of medical device clinical testing can constitute a substantial "barrier to innovation." Bartlett Foote (1991; pp. 81–82) adds that such regulations will have "an inordinate impact on newer, less well-capitalized entrants." As most medical device users would fit into this category it seems exceptionally difficult, especially for higher class devices, for them to pursue such route without the help of a financing, developing or manufacturing partner.

A related problem is that national standards for medical devices have not yet been harmonised.[459] This means that most European medical devices also have to comply with the US standards.[460] The need to go through various separate approval processes creates "an excessive administrative burden" which according to Hutton (1991; p. 150) "may well deter innovation by smaller companies [incl. users] without the resources to meet bureaucratic requirements." For users it is furthermore important that, at least in the US, any new, different or modified claim or indication for use of an existing device will require a separate submission.[461]

In light of the increased importance of nano-, cell-and biotechnology, which is likely to increase the amount of devices in the higher risk classes,[462] the regulatory requirements and the resulting costs and effort of innovation in this area are likely to remain high.

## Product Liability

While product liability is a threat in many if not most US industries, "nowhere, however, does lawsuit abuse inflict more harm than in the arena of medicine" (Hantler 2004; p. 7). Negligence suits can be based on defective design, defective manufacture or failure to warn even if the manufacturer complied with FDA regulations (*Medtronic, Inc. v Lohr*).[463] Lotz (1992; p. 67) explained that "it seems widely recognized that liability laws have had a profound impact on product development in the US." Huge jury awards, including punitive damages—the average US jury award in medical device cases has increased to $ 3.5 million[464]—have substantially increased the legal risk and uncertainty for innovators in this realm.

Various law suits for instance completely reshaped the contraceptives industry by either forcing numerous US companies into insolvency or to withdraw their contraceptives R&D.[465] Japanese silicone makers also seized the production of silicone coating for hypodermic needles that reduce the pain of an injection, out of fear of lawsuits in the US: "We're sure our product is safe, but we don't want to risk a lawsuit" (Hantler 2004; p. 11). Other examples include Union Carbide deciding to forego developing a suitcase-sized kidney dialysis unit or the area of DPT vaccines, where 2 of the 3 manufacturers stopped production in the 1980s due to rising product liability costs.[466] Many types of medical devices are entangled in the "liability environment," such as intrauterine devices, heart valves, pacemakers and anaesthesia equipment to name just a few (Bartlett Foote 1991).

While in Europe this problem is much less severe, medical device product liability laws were also pointed as barriers to innovation in both the Spectaris (2004) and the BMBF (2005) reports in Germany. Substantial medical malpractice risks have the potential to discourage many doctors from modifying, developing and commercializing medical devices.

## Economic Barriers

The expense and time taken to develop medical devices has the potential to decrease user-innovation in this area. Kahn (1991) therefore explains that the majority of newly founded medical device companies fail as they run out of finances. The BMBF (2005) report also reveals a higher failure rate for medical device companies during their first eight years than for other knowledge intensive industries, at least for the investigated time period of 1995–2002.

To illustrate take the example of Dr. Cotes of the University of Cardiff who in collaboration with a university technician developed the "Resparameter." After undertaking various clinical trials he realized that he lacked the financial means to develop the device any further. He therefore contacted a medical equipment company which substantially improved the device, including the addition of various new features.[467]

The cost and effort barrier increases with device complexity. Large companies are therefore generally responsible for innovation in very complex areas such as medical imaging, where the cost of development is especially high.[468] A further important development took place in the US health care financing system in the early 1980s. The introduction of a prospective payment system (PPS) meant that hospitals now received a predetermined fixed amount based on the patients' diagnosis. This system, which is still in place, gave hospitals the incentive to focus on cost-savings procedures.[469] The focus has increasingly shifted to cost-reducing and not quality-enhancing new technologies.[470] We believe that this benefits manufacturer-innovation as these may be more adept in dealing with cost reductions (i.e. process innovations) than users.

## Technological Barriers

Technological complexity and its evolution play an important role in this industry. They constitute a barrier that users have to overcome in one way or the other. In their study of the complexity of the US patent system Allison and Lemley (2001) found that on average the amount of prior art references per medical device patent were 25.84 and therefore far higher than in any other US industry.

Most physicians, while medical experts, are "not likely to know the details of existing technology and [are] even less likely to have the requisite [abilities in] physics or mathematics" (Eden 1984; p. 55). Designing a medical device will furthermore often require substantial engineering know-how and increasingly software-programming skills which most doctors currently lack. The complication of design is illustrated by a FDA medical device recall study in the late 1980s where around 44% of all recalls were attributable to design defects.[471] For user-modifications it is enlightening that "changes made to existing products . . . often produced new defects" (Whitmore 1997; p. 43). In the same way that a typist knows very little about the construction of his word-processor, Eden (1984; p. 59) argues that most physicians will know very little about the construction of for instance an electrocardiograph. Lettl et al. (2006; p. 261) concur: "user-innovators, of course, did not have all the needed competencies and lacked the resources required to transform their concepts into prototypes and marketable products." This "required set of competencies exceeds the 'bounded rationality' of a single user or even a team of users" (Lettl et al. 2006; p. 261).

Some illustrative examples may be useful. Lettl et al. (2006) for instance outline the development of a biocompatible hernia implant. While the University of Aachen in Germany developed a suitable material the leading surgeon at the local University Hospital, Professor Schumpelick—a clinical user with special university resources—hired specialist engineers to develop a camera to verify the specific characteristics of the abdominal wall as he lacked the competencies himself. As Schumpelick also lacked the relevant know-how himself to develop a prototype he hired textile engineers for this process. Finally to complete the finished technology, he collaborated with a leading manufacturer of medical devices to develop and commercialize what was to become the state-of-the-art in hernia repair.

In the various areas such as gynaecological laparoscopy, cochlear implants, nuclear cardiology and echocardiography—two types of sophisticated cardiac imaging technologies—and integrated video-endoscopes users originally played an important role, especially in terms of idea generation and concept formulation.[472] The further development of the technologies however increasingly required close collaboration with teams of specialists that often came from manufacturers.[473] The inherent complexity and multitude of engineering skills required in these areas therefore meant that prototyping and subsequent developments therefore mainly came from industry.[474] The high capital intensity of such machines also played a crucial role. Doctors alone would not have been able to develop such technologies as "their knowledge of the technological possibilities of these imaging modalities was necessarily limited" (Finkelstein et al. 1995; pp. 125–126). In respect of video-endoscopes, Japanese manufacturers, such as Fujinon, Pentax, and Olympus, by taking advantage of their capabilities in video technologies, skills no or very few medical doctors possessed,[475] were able to further improve the technology. By inter alia "incorporating new chip technology into their cameras" (Gelijns and Rosenberg 1995; p. 89) they quickly dominated innovation in this area.

Such control over existing medical imaging technologies can be seen from the inherent difficulty of users to modify such equipment. Not only will the source code to such devices not generally be available to users, but the time, effort and skills needed to enhance the devices will generally be significant. For example take Hewlett Packard's (HP) Color Doppler, a device that uses ultrasound to measure blood velocity. Even relatively minor internal upgrades to a prior version "required a sophisticated understanding of the architecture of the existing systems and input from some of HP's best engineers" (Finkelstein 1995; p. 139).

At a minimum, the increasing sophistication of technology in this realm indicates that user-innovation requires a professional approach. Consider the realm of medical imaging again. For effective innovation to occur clinical expertise needs to be combined with technical expertise in electrical engineering and nuclear physics. Bridging the existing "gulf separating the engineers charged with developing imaging technology from the clinicians charged with interpreting and using the resulting images," which exists due to differences in training, knowledge, experience, and professional orientation—was an "exceptionally challenging" task, that can be overcome only by professionally coordinated collaboration (Finkelstein et al. 1995).

## Social Barriers

Social barriers, operating sublimely, can at times be powerful factors that can both encourage or discourage user-innovation. This can be seen on the basis of early experimentations with cochlear implants. Various early tinkerers, such as Blair Simmons, William House or C. Eyres, all

*Table 4* Summary of Barriers to Medical Device Innovation

| *FACTORS* | *BARRIERS to User-Innovation* |
|---|---|
| ECONOMIC | Expense of Innovation |
| | Long Product Development cycles |
| | Market pressures |
| | Focus on Cost-efficiency |
| SOCIAL | Societal pressure (e.g. stigmatizations) |
| | Organisational pressure |
| POLITICAL/LEGAL | Patent Thickets |
| | Regulatory System |
| | Product liability regime |
| | Political influences |
| TECHNOLOGICAL | Non-modularity of many products |
| | Product design |
| | Interdisciplinarity of R&D |
| | Production/manufacturing |

Source: Own diagram, partly based on BMBF (2005)

decided to halt their experiments, despite some very promising results, due to adverse social pressures. While the lack of suitable materials with which they had to work and the fact that that the technology was still at a very early stage contributed to their decisions, "the recalcitrance of the[ir] communities" seems to have been the main reason for their actions (Blume 1995; p. 102). Major criticisms of their work and the subsequent professional consequences substantially discouraged these experimenters. Take the example of Blair Simmons, whose results were not only rejected by the American Otological Society in 1965, but whose professional funding was also halted by the US National Institutes of Health. He therefore seized to work on "human auditory stimulation" until 1971 when the hostile attitude had changed and he was able to obtain some level of support from his university colleagues (Blume 1995; pp. 102 and 118). An adverse social environment can therefore constitute a significant barrier to user-innovation.

To summarize this section, we can see that many of the barriers discussed previously also apply to the medical devices industry. We will now further illustrate this with a specific example from the German Medical Device Industry: Computerized dentistry.

## COMPUTERIZED DENTISTRY EMPIRICAL ANALYSIS

CDENT is the acronym that we will use to refer to a CAD/CAM (Computer-Aided Design/Manufacturing) system that enables a dentist to design, manufacture and fit tailor-made ceramic restorations without additional laboratory support.[476] The system consists of two separate units, one for imaging and one for milling. Using a computer-driven laser lighted video camera three-dimensional digital images can be captured. The software subsequently determines what the desired shape of the requisite inlays, onlays, crowns or partial crowns should be, which an attached robotic milling device then carves out of ceramic blocks.[477] The original idea that soon resulted in the first functional CDENT prototype were developed by a Professor of dental aesthetics at the dental school of the University of Zurich, a sophisticated user, and a befriended engineer. After a license agreement, Siemens (today Sirona) substantially improved the technology. CDENT was the first commercially available CDENT system in the dental field and is still "the most developed and also most widespread" (Reiss 2006; p. 12). Currently there are over 20,000 systems in clinical use around the world.

CDENT was chosen as a suitable object of study for a variety of reasons. First of all, as far as we know, no other user-innovation study has yet been undertaken with respect to dentists. As existing studies.[478] have indicated that doctors often innovate, we thought this would also apply to dentists. As dentists are not only medically trained, but also skilled craftsmen, many of them also consider themselves to be tinkerers.

More importantly, after having studied how the system was developed and speaking to representatives of the manufacturer, including its CEO, we realized that there was a considerable gap between theory and practice. While user-innovation theory explains how users are increasingly able to innovate and tailor-make products to suits their needs, industry often denies or down-plays such views. In the case of CDENT, the manufacturer, Sirona, expressed the view that user-modifications were not common. Its CEO commented that as we are dealing with "end-users" and not industrial intermediary-users, user-amendments of CDENT were "unlikely." The product is simply "too complex," especially as these users "are not trained engineers . . . [or] software rogrammers."[479]

We therefore wanted to determine whether the view of von Hippel and his colleagues or the corporate/traditionalist view of manufacturer-generated innovations was more accurate in this area. The development of computerised dentistry seems to have been driven and influenced to a large extent by the manufacturer. We believe that users, especially through their continuous feedback, however also played an important role in the subsequent improvement of CDENT.[480] We therefore intended to obtain empirical proof for the existence of barriers to user-innovation, which we believed to be a critical part of the explanation for the discrepancy of the two views.

We specifically chose to investigate the German market for CDENT as the technology was first commercialized by a German company, which for the last 20 years has further improved the technology, and as it is one of the most developed markets with around 10% of all German dental practices employing the technology.[481]

## Methodology

In December 2006 we sent a questionnaire to 1,687 CDENT users. Given approximately 5,000 CDENT users in Germany we contacted around 34% of German CDENT users. We asked respondents whether they had modified CDENT, and if yes in which way. We searched for user-modifications (including third party modifications) which are either changes made to the hardware or software, or new uses of the technology made *independent* of the manufacturer.[482] Any new software update or complementary gadget made or provided by the manufacturer therefore is not a user-modification. For everyone who responded affirmatively, i.e. who claimed that he/she had modified CDENT in some way, we probed deeper to determine whether or not the mentioned act constituted an actual modification. This was done by consulting CDENT experts, the manufacturer, as well as by attempting to get back in touch with the respondent, which was only possible if they had given us a way of identifying them.

Question 16 was the most important part of the questionnaire. It asked the respondent to assess whether the following factors constituted barriers to user-innovation:

1. The patent protection of CDENT can prevent modification.
2. Copyright and the closed source code do not permit modifications.
3. Hacking (unauthorized product modification) is generally not desired.
4. The CDENT technology is too complex to modify, i.e. the user personally lacks the skills to modify a) the hardware and b) the software.
5. The software-license [EULA] forbids modifications.
6. Modification would lead to the loss of the product liability of the manufacturer, breach of warranty and loss of authorisation of CDENT as a medical device.
7. It would be too expensive to modify CDENT or to have it modified by third parties.
8. The effort, in terms of time, to improve CDENT would be too big.

We then asked how the respondents believed their ability to modify CDENT had changed over the years. Whether it had become easier for dentists to modify CDENT as their ability to modify with new technologies had improved, whether it had stayed the same or whether it had decreased as modern dentists may have the ability to utilize new technology but rarely to improve it.

Finally we asked whether the respondent had previously had an idea for an innovative medical device apart from CDENT and to what stage they had developed their idea. The final questions enquired whether the users were content with the interaction with the manufacturer, whether they believed that they would be capable of improving CDENT further in collaboration with the manufacturer and other users, and finally whether they believed that they would be able to improve CDENT if its software were open-source.

## Results

Overall we received 283 responses, leading to a satisfactory response rate of 16.8%. At first a few general words about the survey respondents seem appropriate. We found a high extent of heterogeneity among CDENT users. Different dentists employ the technology in different ways, with different frequencies and with different intentions. While some base their entire clinic on the technology, others use it merely as a useful complement. While some own their personal CDENT, others use it in cooperation with a pool of dentists. Some produce mainly single filling inlays, others multiple ones. This implies a heterogeneity of user needs, which according to prevalent user-innovation theory suggests that a manufacturer will struggle to fully satisfy such need.

Indeed, we found that CDENT users, although generally satisfied by the technology still have a need to improve the technology. The vast majority, 73% to be precise, of respondents indicated that they had some need for improved indications. Forty-eight percent of respondents indicated that they desired further CDENT applications. Users mentioned that they desired improvements in the powdering process (for the digital recording), in the building-up of the occlusal area or in terms of the fitting accuracy. Complaints about the construction suggestions were especially fierce. While some users phrased it politely—“suggested restorations have to generally be adapted considerably”—others were less keen on disguising their frustration: “suggestions are frequently abysmal.” In a telephone interview with the director of the GZFA,[483] the German Society for Dental Health and Aesthetics, we were referred to many of the product’s flaws. The GZFA director emphasized that German dentists employ CDENT mainly for profitability reasons (confirmed empirically by Bühner 1997) rather than to improve the overall quality of the patients’ treatment. The penetration rate of 10% also indicates that the majority of dentists in Germany still do not employ CDENT and generally obtain their teeth from dental laboratories. We take these comments as an indication that there is a considerable leeway for CDENT to improve, especially in the higher-quality market.

Given a need to improve a technology, the development of precise improvement suggestions and the passing on to the manufacturer seems to be a common occurrence.[484] Lüthje (2003) found that 39.2% of responding German surgeons had had an idea for a new surgical product. We consider this to be a very high amount. In our study the solution space was limited to one specific complex technology and therefore much more constrained than the broad area of surgery, where minor modifications in technique are an everyday

*Table 5* Selected Improvement Ideas

| |
|---|
| A construction/design toolkit in which a restoration is possible from the very beginning without the systems' suggestion—needed due to the frequently unsatisfactory construction suggestions |
| To design the manual positioning of crowns in the same ways as the "inlay positioning" |
| A dust cover for the machine |
| The ability to rotate constructions 180% before milling |
| The plastic camera support should be made thicker and softer to ensure that the camera is more stable when recording |
| Manual reset of the water-tank display for extraordinary changes (e.g. in case of vacation or illness) |

surgical occurrence.[485] Unlike surgeons dentists are not generally embedded in a university or hospital environment. Most dentists operate in their own clinic and therefore lack access to both financial resources and the know-how available in a networked setting. The entrepreneurial nature of a dental clinic also means that less time can be devoted to experimentations. Another difference between surgeons and dentists is the different patent scenario. While medical procedures are exempted from patent protection in most countries including the US, and are instead regulated through peer review,[486] medical

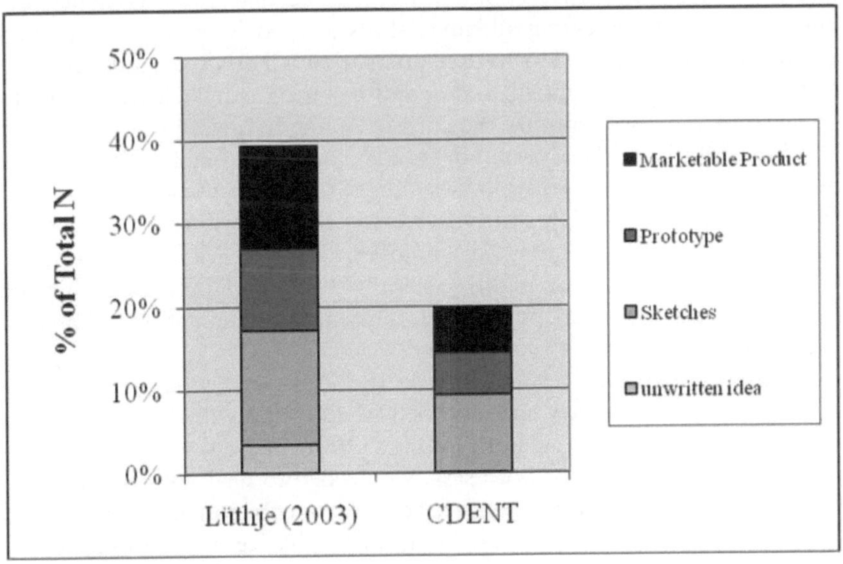


*Diagram 4* Comparison of medical user-innovation.
Source: Own Diagram

devices are frequently subject to patents. This may lead to a considerable difference in the ability to modify a medical device.

Maybe these factors explain why only every fourth respondent (25%) indicated that they had had a (relatively) precise improvement idea for CDENT. A list of some of the ideas can be seen in Table 5. It is noteworthy that this refers simply to ideas and not to any further steps taken in regard to these. Just like Lüthje (2003) we did not further scrutinize these ideas as we did in the case of the modifications. On a closer scrutiny however various improvement ideas would not have merited the term "precise." The actual figure would therefore most likely be considerably lower.

Apart from improvement suggestions we also attempted to determine the dentists' general innovativeness by asking them whether they had had an idea for another innovative medical device and how far they independently proceeded with it. Lüthje (2003) had found that 22% of German surgeons either developed a prototype or a marketable product. While again we made no general efforts to test the validity of the responses, which would have certainly decreased the number of valid responses considerably, we found that of the 20% who asserted that they had had an idea for an innovative medical device, nearly half independently reached a prototype or product stage, which amounts to 10% of the responding population. While Diagram 4 demonstrates that far less dentists had innovative ideas than surgeons, it still illustrates that dentists constitute an innovative group of users.

Subsequently, we inquired as to whether respondents had modified the CDENT device independently or with the help of a third party. We believed that only a small amount of CDENT users will have modified their machines in an innovative manner. We asserted that this was principally due to users not possessing adequate skills to modify the technology and due to the discussed barriers to user-innovation. CDENT is a sophisticated technology and most dentists and dental technicians have learned how to make teeth and not to modify the corresponding hardware.

We found that 67% of respondents indicated that they believed to have no or only a low ability to modify CDENT. To test those who asserted that they possessed a high ability to modify CDENT we assessed whether they had also stated that they had modified CDENT. We found that those who ranked their abilities highly generally modified CDENT much more frequently than the total population. This at least indicates the internal validity of the results.

At a first glance 44 respondents or 16 % of those who answered the question claimed that they had modified CDENT individually or through a third party in some way. As both Lüthje's (2003) and our study largely rely on the subjective assertions of users, we needed to deal with the inherent response bias.[487] Innovative users are generally much more likely to respond than non-innovative users: Users "tend to emphasize or even overemphasize the amount of reinvention that they have accomplished" (Rogers 2003, Rice and Rogers 1980). Not only should such high numbers of innovative users in a self-assessment study always be taken with a pinch of salt, but also be subject to some extent of second testing or triangulation. We therefore made follow-up telephone calls and together with consultations with two CDENT experts reclassified the alleged modifications.

In this process we found that various "yes" answers did not amount to modifications according to our search criteria. For example as a subsequent coloring of a created inlay in the dentist's laboratory is a normal process in the field it also does not refer to a modification of the underlying technology. Similarly, restorations that were changed on the PC to enhance performance as one respondent wrote, are a normal function of CDENT and thus not a user-modification. In the same way the three users who mentioned that they had created so-called dental-neck-veneers/dental-neck inlays were held not to have modified the technology despite the fact that there was no special application for these indications, as the manufacturer specifically indicated that this was a normal process taking 2–3 minutes that its users frequently

*Table 6* The Innovative Potential of CDENT Users

| *Have you had an idea for an innovative medical device outside the realm of CDENT?* | | *To what stage did you independently develop your idea?* | |
|---|---|---|---|
| N= | 270 | N= | 52 |
| Yes | 20% | Sketches/Outline | 48% |
| No | 80% | Prototype | 25% |
| Median | Yes | Marketable product (No patent) | 14% |
| | | Marketable product with patent | 13% |
| | | Median | Sketches/Outline |

| *Selected Non-CDENT User Medical Device Ideas* | *Stage of Independent Development* |
|---|---|
| Oral-hygiene device | Prototype |
| Bridge-remover | Prototype |
| Miniature fastening clamps for the insertion of "Maryland"-type bridges | Prototype |
| Mirror with air-cooling against fogging | Prototype |
| Information-terminal | Marketable product (no patent) |
| A cleaning device to extract mercury from the tooth in order to create bio-compatible implants | Marketable product with patent |
| Implant-robot | Marketable product with patent |

undertook and that it even taught in its training courses. Further "modifications" that were excluded were the use of a third party program for the creation of artificial synthetic dental provisions that the manufacturer had specifically advocated and was offering for download on its website, and the installation of new graphic cards or CPUs that could be ordered from the manufacturer or his suppliers. The creation of "extension-dental-bridges" on the other hand was classified as a modification as these were not specifically authorized by the manufacturer. While it was possible to create these via CDENT, due to potential milling difficulties, the manufacturer did not intend to assume responsibility for these. The dentist in question therefore created and employed these entirely on the basis of his own risk-benefit assessment.

Overall we rejected 26 declared modifications and permitted 18. Table 6 gives an indication of the user-modifications that we encountered to be worthy of that classification. The overall user-modification rate therefore dropped to around 6.6%. Additionally, as we had also asked for third-party modifications we included 3 users who merely used a third-party, non-manufacturer authorized powdering device. This can be justified on the basis that although the use of a third-party powdering device not intended or authorized by the manufacturer they amended the CDENT methodology. Strictly speaking however they did not modify the underlying CDENT technology. If we would exclude these the rate would drop to 5%. Given our conservative attitude in favour of including user-modifications[488] and the response bias that those who actually modified the technology are more likely to reply, the overall percentage of those who are modifying the technology of the total CDENT user-population is likely to be even lower.

The most noteworthy three users were a dentist who together with a German software firm was developing a 'definite' software program to assess the pressure on crowns and bridges in order to find weak points and therefore to lower the risk of the indication breaking; the user who had successfully created extension- and inlay-bridges via CDENT 3D (remember that this was the most sought-after additional indication that respondents had demanded); and the German dentist who developed the PowderChamp.

To briefly outline the PowderChamp, we need to explain that in order for the camera to make an adequate optical impression powder has to be applied to the teeth. Eight percent of those responding to the question indicated that the powdering process needed to be improved. On one chat forum a user stated that "I think the powder is the weakest link in the entire system."[489] A German Dentist, Dr. Horst Landenberger, being "frustrated powdering,"[490] in collaboration with Otopront, designed and developed a powdering device, called PowderChamp, that through air-pressure is able to evenly spread the powder. A doctorate study, which evaluated 15 powdering and spraying devices, found that the PowderChamp was one of the 3 best devices on the market.[491] Interestingly the manufacturer does not consider that powdering is a problem and has shown no interest in the device.

It does have to be acknowledged that though potentially very useful and valuable the majority of the modifications were not in regard to the core components of CDENT. Only one user tinkered with the milling device, only one allegedly modified the imaging unit and the core software remained completely untouched. While 2 rewrote the system.ini file this did not change the software as such, just their particular ability to utilize it effectively for their purposes.

Modification is in any case merely the re-invention stage of innovation. Diffusion of the modification remains a critical component of user-innovation. In our interviews we discovered that only Dr. Landenberger had taken this step of commercializing his development, with limited success so far. The "definite" software may also be commercialized in the near future. Overall we can therefore summarize that the vast majority of improvements in this realm were controlled, if not made, by the manufacturer. User-innovation in regard to CDENT seems to have been marginal.

The next step was to determine why over 90% of users did not modify their machines when the majority had a need for improvements? The answer to this question is that certain barriers to user-innovation exist that burden user-modifications. The overall median response for all the barriers was 3, standing for "high barriers." As it is the collusion of the individual barriers that really matters this finding is noteworthy. This high median for the individual barriers therefore indicates that taken as a whole users consider the barriers to user-innovation in the realm of computerized dentistry to be "high." The majority of users, 53% to be exact, chose this option. In 6

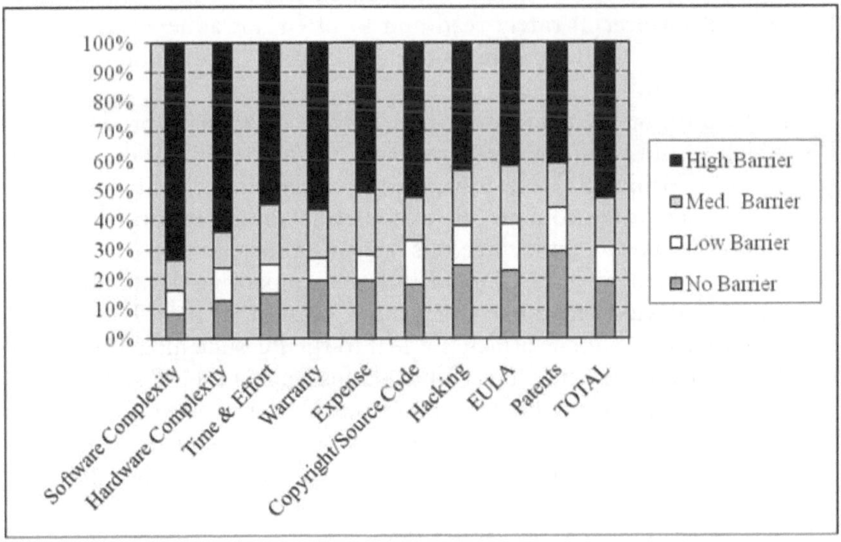


*Diagram 5* Barriers to CDENT user-innovation.
Source: Own Diagram

of the 9 areas the high-barrier option was chosen by over 50% of the users. "Hardware Complexity" was held to be a high barrier by nearly 2/3rds of respondents. The option "software complexity" was considered to be a high barrier by nearly 3/4ths. Around 70% of all users classified the nine factors as medium or high barriers to user-innovation.

We ranked the different barriers and found substantial differences between them. Table 7 summarizes this. While hardware and especially software complexity were considered to be very high barriers, patents, hacking and restrictive software licenses received far less support. Nevertheless even these with medians of 2 and averages of 1.67, 1.8 and 1.8 respectively they were still considered to be moderate barriers. The distinction touches upon the important difference between ex-ante and ex-post consequences of the barriers. The issue of complexity, just as the time and effort needed to innovate which was also considered to be a high barrier, operate ex-ante or before the user can improve the technology. Skills and resources are needed to circumvent the barrier. Patent infringement, breach of a software license or the social stigmatization of hacking operate mainly after the user has already made the improvement. While they can prevent the commercialization or further pursuit of the development, if detected, they previously merely serve as a psychological barrier, which is easier to overcome than a tangible one.

Some comments in respect of the individual barriers seem appropriate. To start with the legal barriers, the 135 CDENT patents that the manufacturer held globally as well as CDENT's standard non-modification clause, were merely held to be moderate barriers to user-innovation. As the widespread copying and exchanging of copyrighted music for instance indicates, EULAs are however a) rarely read and b) often not adhered to. In both cases, the barriers only operate ex-post, after the actual modification has already occurred.

In regard to the regulatory barrier, CDENT is a Class II product in the US, which would require the assurance that it meets "applicable industry standards" (Termini 2004; p. 116). As we believed that the likelihood was low that a user would commercialize a CDENT modification without the manufacturer, we did not include this factor in the questionnaire.

The risk of breaching the CDENT warranty and losing the authorization of a medical device was ranked as a more substantial risk. The CDENT manufacturer for example attaches a seal to the PC slide model that states that "if the seal is broken, all warranty claims against the PC slide-in module are invalid" (section 8.4 of the Application Unit Instructions). To further protect itself from legal liability for potential injuries incurred by modified machines, CD's 3D Hardware Compliance Requirements, states that

> the use of any hardware, other than what is supplied by Sirona, will violate the compliance of the CDENT 3 System and compromise its safety and efficacy. *The use of non-approved hardware will void the*

*warranty and support from Sirona Dental Systems.* (Sirona 2004) [Own emphasis added]

A modification of CDENT could lead to the loss of CDENT as an authorized medical device. If a patient suffers damage as a result of an unauthorized modification a malpractice claim is likely to result. Various courts, especially in the US, have held doctors liable for harm resulting from the experimentation with novel unproven procedures or technologies.[492] Although the legal environment is less virulent in Germany, an unauthorized modification will still be a risky undertaking for a doctor.

Both the expense and the time and effort of modifying CDENT were found to be high barriers to modification. We are dealing with a sophisticated, non-modular technology that the manufacturer does not want to have modified by third parties.[493] The time demands on the dentist in the general course of his clinic (indicated by comments through the telephone interviews and the difficulty of getting in touch with dentists in the first place) and the time it took to master the technology (given that the average respondent had been using the technology for over seven years and that around 50 % of users had attended an expert course it is surprising that over 60% acknowledged that they still did not fully master the various indications of the technology) strongly indicate that the majority of dentists will not attempt to improve the main functionalities of their machines. Many users realized that they may not be in the best position to improve the technology:

- "too burdensome, no time, no desire, I am a user"
- "I am a dentist, not an engineer!"
- "It is nonsense to modify the machine yourself"
- "I simply don't have the time to experiment"
- "the effort [and cost] involved in tinkering has to lead to a net-gain, which I can not imagine it does"

Through our research we found various indications for the assertion that the company strives for complete "control over the product and its uses" as one user explained. Sirona therefore allegedly exerts control not only over how the product is specifically used, but also over all training courses and to some extent over the most important publications in this realm. A considerable amount of market pressure is therefore exercised, especially given the dominating market position of the manufacturer. As this factor only evolved as an important one once we analyzed the responses, we did not include it on the questionnaire. It remains an important factor for future investigations.

The technological barriers were found to be the most substantial barriers. Users even complained about the whole workings of the technology being kept "secret," which is however common industrial practice in most areas. Over the past 20 years the interaction between the software and the hardware has been steadily optimized, and while there may be room for further improvement, it will be difficult for outsiders not possessing the accumulated know-how of the manufacturer to enhance the system. The closed source-code and CDENT's sealed and non-modular design are clear examples of the manufacturer's non-modification policy. The instruction manuals for both the application and the milling unit of CDENT 3 for example state that

> [f]or safety reasons, the unit may only be operated with original Sirona accessories or other accessories which have been approved by Sirona. The owner assumes the risk for the use of non-approved accessories. . . . *Opening of the unit with tools is strictly prohibited.* [Own emphasis added]

Such statements indicate that the manufacturer aims to make all improvements himself. A further technological measure that is intended to prevent unauthorized use is the 'dongle.' This refers to a cable that has to be plugged into the computer to authenticate the hardware. Dongles are similar to the trusted computing concept discussed earlier and are intended to protect expensive software packages. While CDENT's dongle does not prevent the use of the software itself, without it the milling unit will not function. While there are ways to outsmart this copy-prevention mechanism, it certainly increases the difficulty and effort involved in tinkering.

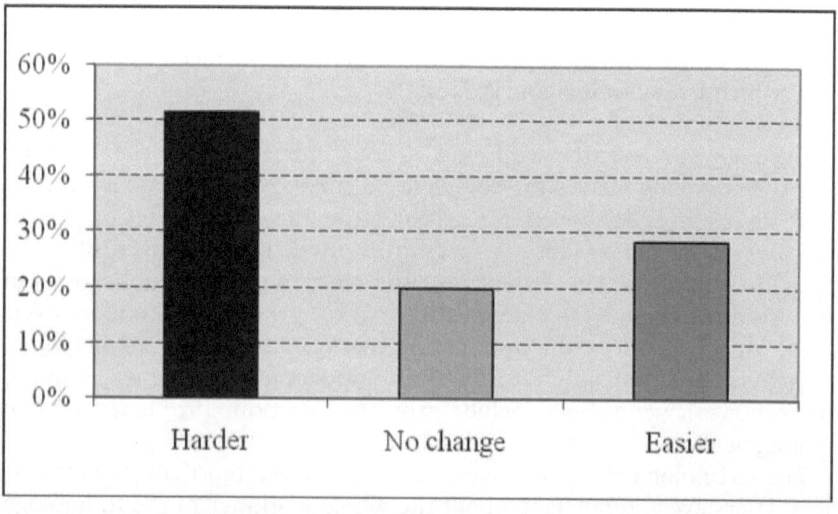


*Diagram 6* The progression of barriers to user-innovation.
Source: Own Diagram

*Table 7* Barriers to CDENT User-Innovation

| | *1. Patents* | *2. Copyright / Closed Source Code* | *3. Hacking* | *4a. Hardware Complexity* | *4b. Software Complexity* | *5. EULA/ Software License* | *6. Warranty/ Medical Device Authorization* | *7. Expense* | *8. Time & Effort* | TOTAL |
|---|---|---|---|---|---|---|---|---|---|---|
| N= | 209 | 206 | 199 | 189 | 184 | 201 | 209 | 205 | 210 | 1812 |
| No Barrier | 29% | 18% | 25% | 13% | 8% | 23% | 20% | 20% | 15% | 19% |
| Low Barrier | 15% | 15% | 14% | 11% | 8% | 16% | 8% | 9% | 10% | 12% |
| Med. Barrier | 15% | 15% | 19% | 12% | 10% | 19% | 16% | 21% | 20% | 16% |
| High Barrier | 41% | 52% | 43% | 64% | 73% | 42% | 57% | 51% | 55% | 53% |
| Average | 1.7 | 2.0 | 1.8 | 2.3 | 2.5 | 1.8 | 2.1 | 2.0 | 2.14 | 2.0 |
| Median | 2 | 3 | 2 | 3 | 3 | 2 | 3 | 3 | 3 | 3 |

0= No Barrier
1 = Low Barrier
2= Medium Barrier
3= High Barrier

Source: Own Table

*Table 8* Ranking of Barriers to CDENT User-Innovation

| | *Average* | *% High Barrier* | *Actual Barrier Classification* |
|---|---|---|---|
| 1. Software complexity | 2.5 | 73 | Very High |
| 2. Hardware complexity | 2.3 | 64 | Very High |
| 3 Time & Effort | 2.1 | 55 | High |
| 4. Loss of Warranty/Medical Device Authorization | 2.1 | 57 | High |
| 5. Expense of Modification | 2.0 | 51 | High |
| 6. Copyright/Closed Source-Code | 2.0 | 52 | High |
| 7. Hacking | 1.8 | 43 | Medium |
| 8. EULA/Software license | 1.8 | 42 | Medium |
| 9. Patents | 1.7 | 41 | Medium |

The manufacturer clearly indicated that no user tinkering is desired. Indications for the existence of organizational pressure on users not to tinker, by attempts to discredit such outlaw users were also found. Two respondents stated that the manufacturer "blocks" improvements. Another comment was that "Sirona is monopolising the market." While this amounted only to a moderate barrier it still indicates that various users were put off by such pressure.

To understand the dynamic evolvement of the barriers to user-innovation in this field, we asked respondents whether they found the recent version of CDENT easier or more difficult to amend than the prior version. We found that over half of the respondents believed that it had become more difficult to amend CDENT in comparison to prior versions (28% believed that it had become easier). This indicates that while modern dentists may be more acquainted with computers and high technology, it seems that the amount of know-how needed to improve CDENT has increased over the years. There seems to be a discrepancy between the skills and resources needed to innovate and those possessed by users. Few dentists will learn how to engineer medical device or how to program code. As one user phrased it: "I don't have a clue about programming." This may well change however in the forthcoming years if programming skills become more common in society as a whole.

The closed source code, especially in its inherent relation with the software complexity, was considered to be an especially high barrier for the respondents. We therefore inquired how many CDENT users believed that they would be able to tailor-make or improve the CDENT software if it were open-source. Around 41% of respondents or 108 users answered that they considered themselves capable of this. To further investigate this we asked whether these users believed that software complexity was a barrier to user-innovation. We compared the percentage of actually made modifications of those who indicated that they believed that they could improve an OSS version of CDENT and those who did not. We found that while 10% of the OSS enthusiasts had previously modified CDENT, this applied to only 4% of those who believed that they would not be able to program CDENT to suit their needs. This lends support to the view that if the source-code were open there would be some users who would most probably be able to tailor-make the software to their needs and to potentially even improve it.

Thirty percent of users seemed to be dissatisfied with the interaction with the manufacturer. Many complained about the manufacturer's hotline. One dentist even referred to the hotline as "incompetent". This is important as users strongly believed that they were able to further improve the technology in case of a closer cooperation with the manufacturer. Seventy-six percent of users choose this option. While this arguably amounts to an overestimation of their own personal abilities and resources, it certainly indicates the desire of users to help improve the technology. What we found even more interesting was that the dissatisfied users appeared to be more valuable as cooperation partners. We found that not only did they have an even stronger belief of their ability to help improve the technology than the general user-population, but they also modified the technology to a greater extent. This was especially the case with the verified user-modifications were dissatisfied users modified 102% more than satisfied ones. Table 9 illustrate this. Two further dissatisfied users were found that were not part of the survey that had also used the software in intriguing and novel ways. Mollick (2004) in his work on "parasitic innovation" already pointed out that it may be worthwhile to collaborate with outlaw users. Currently the manufacturer mainly cooperates with

*Table 9* The Innovative Potential of Dissatisfied Users

| | *General population* | *Dissatisfied users* | *Δ** | *Δ in %* |
|---|---|---|---|---|
| N= | 274 | 74 | | |
| Believe that they could modify | 76% | 88% | +12 | + 16% |
| Do not believe | 24% | 12% | - 12 | - 49% |
| % Self-assessed high ability to modify CDENT | 12% | 19% | + 7 | + 60% |
| Average self-assessed skill-set | 1.25 | 1.33 | + 0.1 | + 6% |
| % Modification (self-declared) | 16% | 18% | + 2 | + 12% |
| % Modification (verified) | 7% | 13% | + 6 | + 102% |

* Percentages refer to exact figures, not the rounded up figures

so-called Beta-users, which provide critical feedback to the manufacturer. The manufacturer should attempt to integrate highly unsatisfied users in some way into their innovative efforts. Additionally, it may be worth establishing a hotline or a suitable website to specifically handle users' ideas and modifications to remedy the current dissatisfaction.

## CONCLUSIONS AND EFFECTS

The important role that users play an in the development of technology, cannot be reemphasized enough. They not only frequently use products in novel ways but also occasionally manage to significantly improve these. Many industries will therefore be characterized by both manufacturers and users contributing to the innovation-generation process. In this chapter we wanted to demonstrate however that users may not always be able to innovate due to the existence of various barriers. In an exploratory manner, we saw that various types of users from security researchers to dentists have experienced how market power, legal and technological developments and social stigmatizations can collude to impose substantial barriers to user-innovation. As Burk (2004; pp. 1564–1565) explains, "when combined with statutory enactments . . . or judicial decisions favouring the licensing terms promulgated . . . by [the content industry or the seed developers], these [technological] developments dramatically alter the balance of ownership and control of digital technologies," to the extent that "rights holders can effectively write their own intellectual property statutes."[494] We thus argued that free user-innovation, i.e. without obtaining some type of permission, may frequently be difficult as users often lack not only the legal and contractual rights, but also the technological ability. We saw that some barriers inherently exist, while others are artificially erected by manufacturers. We clearly show that manufacturers can actively change the cost and benefit expectations of users.

In respect of a small niche market, we empirically demonstrated the existence of high barriers to innovate, which prevented users from acting on their need to modify the technology in question. We need to point out however that despite the fact that we tested the asserted user-modifications the results of the study are still largely dependent on the subjective responses of the participants. As the study of the barriers to user-innovation is still in its very beginnings, subsequent research in this area, such as through in-depth case studies or further quantitative empirical investigations, in respect of both existing products and novel technological developments is needed and strongly encouraged. Especially the difference between industrial users and consumers merits further study.[495] We believe that without scrutinizing the dynamics of such inhibiting factors, the phenomenon of user-innovation can not fully be understood.

To summarize our arguments the framework proposed by Lettl et al. (2006) seems useful. They offer a 'swell model of development contributions by users for radical innovation' in which users need specific characteristics to reach a specific level of user activity. In this model, users require i) motivation caused by a current problem, ii) openness to new technologies, and iii) imagination capabilities. To reach the subsequent level of development, where users are able to actively contribute to the user's domain, 4 further criteria are required: a) high competence in their own domain, i.e. a "profound understanding of the elements, the causes and effects of a certain domain," b) tolerance for ambiguity, i.e. the ability to handle uncertainty, c) access to technological know-how and d) resources for their own research activities. Finally, to reach the highest activity level, users also need technological expertise in the relevant technological domains. In other words, if a user, individually or as part of a team, is able to meet those criteria, radical user-innovation would be feasible. Lettl et al. (2006; p. 262) clarify that especially the last criteria is very hard to meet: "Most surgeons do not have the time and the 'absorptive capacity' to build up such technological competencies." In other words, the higher that we move up this swell model, the fewer users are able to meet the requisite criteria.

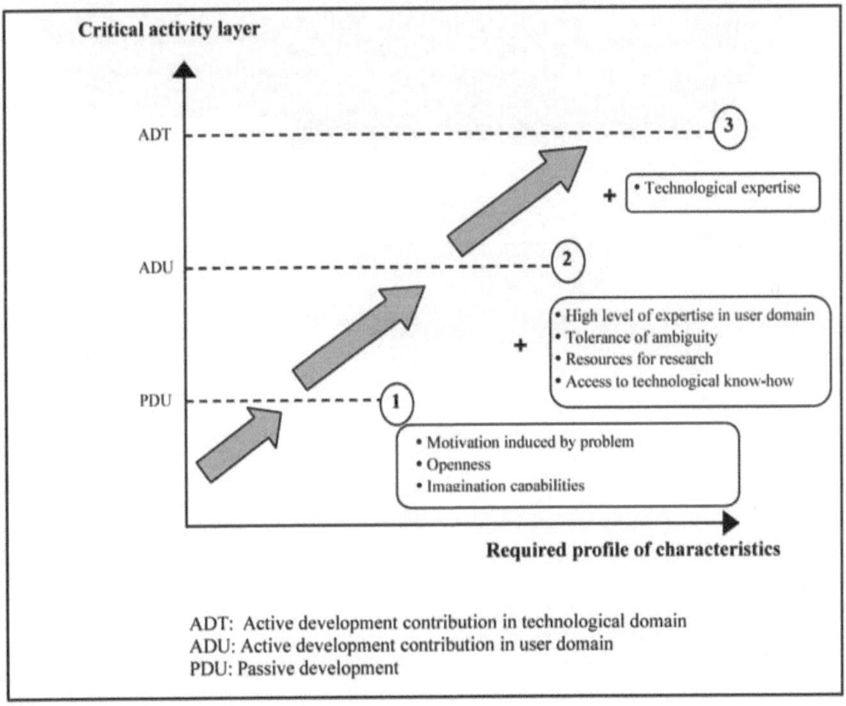


*Diagram 7* Lettle et al.'s swell model for radical user-innovation.
Source: With permission from Lettl, Herstatt and Gemünden (2006)

Three amendments to this model are necessary for our purposes. First of all the model seems to some extent to be applicable to all types of user-innovation, not just radical ones. Secondly, it should be considered that radical user-innovation will require slightly different skills than incremental user-innovation. While for both types of innovations business and technological acumen will be necessary, although to different extents,[496] both the "tolerance of ambiguity" and the "openness to new technologies" are far less important for incremental user-innovation. A third important amendment is the need to take external factors into consideration. Currently, the model focuses exclusively on internal factors. Patent barriers or a hostile economic/social environment will however exist for both incremental and radical innovation in a variety of industries and will have to be dealt with. We therefore add a further hurdle, which technologically advanced users have to overcome in order to become innovators: The ability to overcome external barriers to user-innovation. Innovators in the Independent User-innovation (IUI) phase further distinguish themselves from prior stage by possessing sufficient marketing and manufacturing skills to

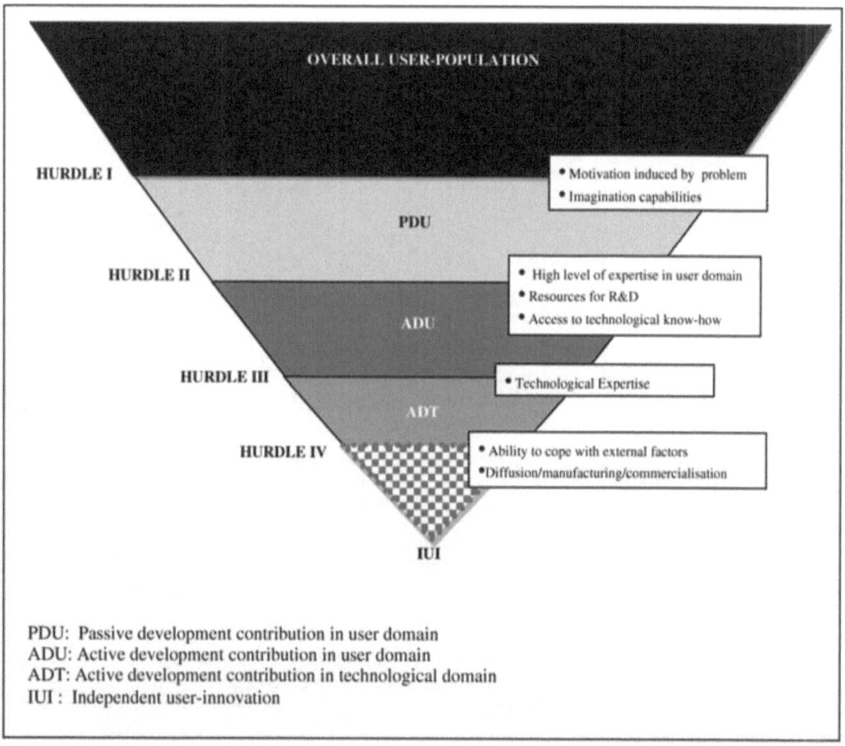


*Diagram 8* Generic swell model for user-innovation.
Source: Own creation, based on Lettl, Herstatt and Gemünden (2006)

enable adequate diffusion/commercialization. This phase is akin to Foxall and Johnston's (1987) UII4, the vertically integrated user-initiated product innovation, and Voss' (1985) user developed category, in both of which the user effectively takes on the manufacturers' role and creates and commercializes the product in question.

In our amended swell model we thus depict a down-ward pointed, inverse pyramid of users of a specific technology that need to overcome certain barriers in order to reach the next stage of innovation. In other words, the barriers seem to function somewhat akin to a stage-gate model,[497] where certain projects, in this case users, can proceed, while the majority are unable to overcome the hurdles to reach the next stage. In this model users, not by the choice of management, but rather by a process akin to natural selection can proceed to the next stage, if they possess the listed characteristics. The further we move down the pyramid the further we leave the MAP and move towards independent user-innovation. Only few will possess the requisite ability and right to innovate. Innovating is harder than merely modifying which in turn requires more skill than mere use. Clearly the case of the independent user-innovator entrepreneur will be relatively rare in most mature industries, and therefore is depicted by the small checkered triangle at the bottom of the pyramid. Lüthje and Herstatt (2004; p. 559) appropriately summarized this assertion:

> Even in markets that are characterized by a high level of user-innovation, only a fraction of the user population engages in development efforts whereas the majority of users remain passive. Improving existing market offers or conceiving new problem solutions is located outside the range of ordinary customer activities.

Our empirical study provides support for this assertion. As Hienerth's (2006) study of the rodeo-kayak industry shows such IUI is likely to be much more prevalent in the early stages of industry.

This is a different take on the democratization of user-innovation argument. Demos refers to the people and democracy to the government of the people. We present considerable indications that make us doubt whether the average user today is more likely to be able to innovate than was previously the case. On the contrary provided evidence for the assertion that independent user-innovation faces a wide variety of barriers which relatively few users will overcome. Numerous factors, such as the industry and the product involved, will influence the extent of these barriers. In the following part, the dynamics of these factors will be further scrutinized.

A final important addition to the democratization debate is that portraying innovation as a bipartite system in which users and manufacturers interact to various degrees is misleading. "Open Innovation" involves the integration of a wide variety of parties such as research institutions, governmental agencies, suppliers, universities, etc. into the innovation

*Table 10* Non-exhaustive Checklist of Barriers of User-Innovation

| **Legal** | **EXAMPLES** |
|---|---|
| Patent Thicket | in semiconductors, biotechnology and nanotechnology. |
| Laws forbidding the circumvention of technological means of protecting copyright | DMCA, EUCD, Australia's Digital Agenda Act 2000 |
| Anti-tinkering and -hacking laws | DMCA, US Computer Fraud and Abuse Act |
| Strict Copyright Laws | EUCD, CTEA |
| Restrictive Database Rights | EU Database Directive |
| Product Liability laws | Medical Devices |
| **Market** | |
| Strict Contractual terms forbidding modification | EULAs, Seed-licenses (II.2) |
| Companies using their market power to increase their control over needed information and know-how | Microsoft, Monsanto (II.2), Petrol companies |
| Closed, non-free or restrictive Standards | Intel, mobile phones |
| Organizational Barriers | Lack of finances, marketing skills, time etc. |
| **Technological** | |
| Product Complexity | a)Lack of public availability of needed know-how<br>b) lack of availability of needed tools |
| Technologies to restrict access to know-how and to prevent user-modification | Trusted Computing, tamper-resistant software, Terminator-technology (II.2) |
| **Social** | |
| Stigmatisation/Labelling | Hackers, downloaders |
| Resistance to Change by Established Players | Edison, Ford |

creation process.[498] While it would thus be misleading to speak about the democratization of user-innovation, arguing that the increased technological sophistication, complexity together with modern business practices has led to an increased amount of players shaping the innovation cycle, seems sensible.

# Part C

# The Dynamics of User-Innovation

The preceding chapter provided numerous indications that the current user-innovation literature is incomplete. A suitable explanation for the difference between the amount of user-innovation in different industries was previously missing. We therefore attempted to fill this vacuum by suggesting that barriers exist that can burden user-innovation. These barriers will obviously differ in different areas. Until now the vast majority of user-innovation studies have merely scrutinized a particular industry at one moment in time and found users to be an important source of innovation. In this respect user-innovation was frequently impliedly treated as if it were a static construct. The few studies which explored how user-innovation changes over time—Hienerth 2006, Riggs and von Hippel 1993, Knight 1963—have however demonstrated the dynamic nature of this process. Our own study provides further support for this assertion. While the idea and first prototype to CDENT originated at least partially from a user, subsequent developments were largely made by the manufacturer. In this part we will build on the previous one by providing some additional foundations for this allegation that the degree of user-innovation will change over time as the barriers to user-innovation and other relevant factors evolve. We will commence with a general discussion of the dynamic nature of user-innovation. An in-depth study of the evolvement of the seed industry with consideration of the barriers to user-innovation will follow. Finally, a summary of relevant influencing factors will be made.

## THEORY & PAST STUDIES

"Capitalism . . . is by nature a form or method of economic change and not only never is but never can be stationary" (Schumpeter 1968; p. 27). If the overall system is dynamic, a logical corollary seems to be that this is due to the constant interaction of its components. Technological evolvements are one of the key drivers of economic changes. While innovation can reshape business structures, it is itself "conditioned by a firm's competitive environment, technology, organization, and conscious strategic choices" (Utterback 1994; pp. vii–viii). The extent of its re-invention will thus depend on a wide variety of internal and external factors such as the different needs, problems abilities, experiences, resources and restraints of users and will therefore change over time.

In the mid 70s Abernathy and Utterback introduced a model of the dynamics of innovation, which remains useful in explaining the type of innovation over the development of an industry. The model describes the extent of product and process innovation in three stages. The fluid phase describes the formative years of an industry in which the rate of product innovation is the highest and the rate of process innovation is very low. In the subsequent transitional phase the rate of major product innovation slows down and a dominant product design evolves which leads to an increase in the rate of process innovations. Empirical evidence in areas such as typewriters, cars, television, electronic calculators, etc. has demonstrated that the number of firms in an industry steadily declines after a dominant design has evolved until only a few large firms remain. Finally, some but not all industries enter the specific phase where the rate of both process and product innovation decreases rapidly and where these markets become commoditized, i.e. focused on cost, volume and capacity.[1]

Nagel (1993: p. 53) in his study of lead-users in the different phases of development of the electronic circuit board industry in Europe and the US demonstrated that the "natural locus of innovation" changed over time during the different phases of the innovation cycle. Hienerth (2006) and Baldwin et al. (2006) in their analysis of the commercialization of user-innovations in the rodeo kayak industry also demonstrated that changes throughout the different phases affect user-innovation. In this case in the mid 1970s, lead-users, to meet their idiosyncratic needs, started to develop and construct specialized kayaks and related equipment to perform acrobatic tricks in rough white water. The free exchange of designs and know-how among users quickly led user-innovation to flourish. As main-stream users increasingly started to appreciate rodeo-kayaks and as these became gradually more sophisticated, various users started to commercially manufacture kayaks causing their motives to change. Rivalry became more intense with the result that "in some cases free assistance was only given to selected community members on a quid pro quo basis" (Hienerth 2006; p. 282). As the industry grew, the traditional kayak manufacturers also increasingly entered the market. Through the 1990s this commercialization trend progressed until "information was only shared freely in the process for non-strategic, non-significant issues," and extreme paddlers increasingly started to purchase their equipment (Hienerth 2006; p. 284).

While it had been relatively cheap and easy for users to modify and manipulate fiberglass boats in the late 1980s and early 1990s, tinkering with plastic boats required a much higher capital expenditure.[1] Consequently, after 25–30 years of users having been the dominant innovators in this field, once a dominant design, center buoyant plastic boats, evolved, traditional manufacturers soon began to control the industry with their capital intensive production systems and incremental innovations. Hienerth (2006; p. 290) summarized: "Communities will change dynamically over time, influenced by the level of participation and the motivation of individual members."

Further anecdotal evidence indicates that in many sports the initially high levels of user-innovation[2] rapidly declines as the industries mature. In various realms from tennis to motor-sports as technology matured we experienced the appearance of a role distinction between users and manufacturers. This is as the conditions that enabled such innovation in the first place, such as "low-cost production (high variable costs but no upfront capital costs), relatively low forms of technology used, high rates of interaction between community members, exchange of skills and know-how" (Hienerth 2006; p. 289), change, so does the ability of users to innovate. In regards to the medical imagining industry, Finkelstein et al. (1995; p. 141) stated that "many of our respondents claimed [that] the activities that make up the innovation process are harder to conduct now than they were in the past."

Such dynamic changes are indeed not only common but inherent in our economic system.[3] A study by Riggs and von Hippel (1994) in the realm of two types of scientific instruments further indicates this: While users were responsible for the vast majority of innovations from the mid 1950s to the late 1960s, from 1969 onwards manufacturers generated the majority of innovations. The software industry is another prime example of how the dynamics of user-innovation shift over time. The industry was largely user-initiated through tinkerers/programmers in academic and governmental laboratories developing software largely to meet their own needs.[4] Early lead-users, even if they worked for different research centers, actively shared and exchanged their ideas and developments.[5] Copying and modifying was not only permitted but encouraged.[6] As distribution beyond specific geographical hubs was still difficult as the internet had not yet been invented, innovation largely occurred in hubs such as CERN in Switzerland or MIT's Artificial Intelligence lab.[7] As in the early 1980s however an increased amount of commercial firms saw the opportunities of commercializing software, closed proprietary software packages "that could not be studied or modified by others"[8] (von Hippel 2005a; p. 98) became increasingly prevalent. While there is no reliable data in this field, it is certain that percentage-wise the amount of user-innovation decreased. As a counter-movement to this trend the free and open software movements originated.[9] Jointly with the increasing availability of computers and internet connections the spread of open software, including various programming toolkits for writing special kinds of code, gave many users access to the tools and know-how needed to modify and improve upon software. User-innovation therefore was able to flourish again. The developments in the realm of software patents,[10] trusted system and software complexity have the potential however to adversely affect this again.

User-innovation should therefore clearly be seen as "a dynamic process best depicted by a continuum" (Foxall and Johnston 1987; p. 93) or a "process of evolutionary development" (Rothwell and Gardiner 1985; p. 168) that consists of invention, innovation and re-innovation. We would like to expand on these works, by scrutinizing the evolution of the seed-breeding industry from being a user-dominated means of survival to becoming big business controlled by suppliers.

## FARMER-INNOVATION IN REGARD TO SEEDS

Farmers, or users of seeds, have historically not only been the worlds' largest group of professional users[11] but also most probably the biggest group of user-innovators. While continuously declining in numbers in the western world,[12] their importance continues to be considerable, especially in developing countries.[13]

To begin with agriculture refers to the systematic cultivation and harvesting of crops. It should be seen as a "cultural activity in which human beings interfere with nature" (Lammerts van Bueren 2002; p. 25). While for many years the general attitude of scientific researchers has been that farmers, especially in the developing countries, do not innovate, since the early 1970s researchers have gradually recognized the substantial role that farmers have historically held and can still play as innovators and experimenters.[14] Through continuous experimentation and adaptation farmers around the world have created a wide variety of innovations ranging from the development of new plowing instruments to natural non-toxic pesticides to enhanced irrigation methods, soil conservation, intercropping or the cultivation of native species which are very well suited for local conditions.[15]

We will focus on one of the most important aspects of farming, namely on the cultivation of seeds. The (i) environmental specificity and (ii) the central role of biological processes (random mutations) distinguish this area from most others.[16] While humans and their predecessors existed for over 2 million years as hunters and gatherers, only around 12,000 years ago did they start to not only harvest but also to collect and sow seeds.[17] The process of seed-cultivation, also referred to as plant breeding, consists out of the following four principal activities.[18]

1. Selection of source germplasm (i.e. saving the suitable seeds)
2. Trait development
3. Cultivar development
4. Varietal evaluation

To get from plant-breeding to seed-innovation diffusion is also necessary. This is an intricate issue as we are dealing with a finished natural product, the seed, that can however be used to create something else, such as a crop. Seed-innovation refers to improving and diffusing seeds, not merely to growing crops in novel or different ways. As in the beginning of farming the systematic exchange of seeds was lacking and was only developed later,[19] in the early days the diffusion criteria has to be seen in a looser manner than today, with any form of propagation, including the local exchange of the seeds, being sufficient.

It was a random process at first: "People selected the tastiest and biggest wild fruits, the seeds from which then started to grow in the rubbish tips and latrines around their camps" (Douthwaite 2002; p. 166). While at first nature's random mutations and adaptations was the sole cause of crop improvements, as time went by people gradually learned to take advantage of nature's dynamics by starting to systematically select seeds. By around 2,500 BC farmers had domesticated (which led these to lose their own defences, such as the bitterness of cabbage) and brought under cultivation "all the major and minor food crops on which humankind survives today" (Rhoades 1989; p. 4).

Since the beginning of land cultivation, seed-users/farmers have therefore been responsible for seed-innovation by selectively saving, planting, breeding and exchanging seeds.[20] Kästler (2005; p. 10) clarifies that "since the beginning of agriculture, the selection and reproduction of seeds, as well as the conservation and renewing of agricultural biodiversity, have never left farmer's fields." The existence of over 200,000 known varieties of wheat or over 100,000 different types of rice that existed before the

*Table 11* Traditional Seed-Innovation

| *Activity* | *Farmer* | *Supplies* |
|---|---|---|
| 1. Selection of source germplasm | | |
| 2. Trait development | | |
| 3. Cultivar development | | |
| 4. Varietal evaluation | | |
| 5. Seed Diffusion | | |

Source: Own Table

= Party engages in designated activity

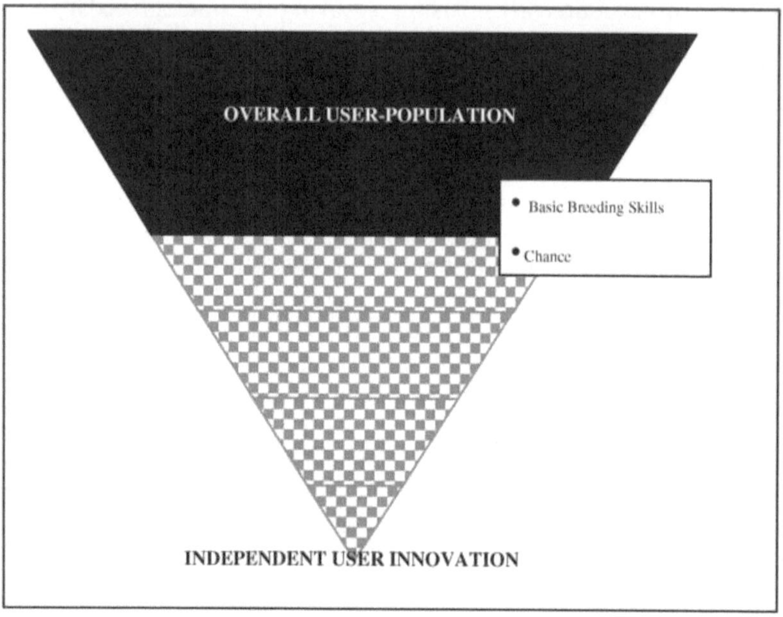


*Diagram 9* Traditional plant breeding swell model.
Source: Own diagram.

Green Revolution evidence the extraordinary success of traditional farmers in developing innovations.[21]

In terms of our swell model the threshold to innovate in the realm of seeds was therefore initially relatively low and many farmers bred improved seeds to subsequently exchange these. Some basic skills in selecting seeds with superior characteristics and luck were the main differentiating factors between innovative and non-innovative farmers.

Although specialized commercial seed traders had been around for centuries, the first experimental breeding stations, where promising plants were systematically grown and evaluated, started to develop in England and Germany only in the early 19th century.[22] In 1862 the US Department of Agriculture (USDA) was formed, which as time went by sponsored an increasing amount of seed research. The result of such systematic breeding were the developments of new varieties, such as the wheat varieties Sherriff and Vilmorin, that gradually began to replace the previously existing landraces, i.e. those early cultivated crop species that consist out of heterogeneous mixtures of genotypes that have stably adapted themselves to their environment.[23] At the end of the 19th century US agriculture was dominated by openly-pollinated varieties which farmers had largely saved and cultivated themselves.[24] The breeding activities of public institutions such as the USDA did however gradually increase in importance. Swann Harding (1947) in his history of the USDA even describes that without the breeding of new varieties by the USDA both the cane- the beet-sugar industries would have practically been destroyed by virus plagues. Soon afterwards disease-resistant varieties from corn to tomatoes to asparagus that USDA researchers had bred started to spread across the US.

During the early 20th century seed-suppliers remained small, family owned businesses. They lacked the financial resources necessary to pursue their own R&D and therefore sold varieties developed in the public domain.[25] During this time "the furnishing of new seed was [still] half farmer skill and half trader alchemy" (Fowler and Mooney 1990; p. 120). As an increasing amount of seed-breeding companies were formed both in the US and the EU, agriculture, from being a way of life, gradually turned into a form of business that focused on profitability and embraced new technology to meet that goal.[26] The goal of these corporations obviously was to breed better seeds than the ones cultivated by farmers to win these as customers.

The first major breeding success that led to a substantial decline of farmer-innovation was the development of hybrid corn by the USDA and private seed companies. Its introduction in the US and beyond in the 1920s and 1930s revolutionized agriculture. Although hybrids enabled farmers to obtain significantly higher yields than regular varieties (so-called heterosis or hybrid vigour), they also, for the first time, made seed saving—a prerequisite for farmer seed-innovation—redundant. As inbreeding hybrids generally leads to genetic weakness and thus to significantly lower yields in successive generations, re-inventing such seeds became a "biological impossibility" (Agarwala-Rogers 1978; p. 140). Therefore hybrids effectively obliged farmers to buy new seeds each year and thus "prevents farmer seed practices of saving, exchange and plant breeding that are foundation of agricultural systems" (GRAIN 2005a; p. 13). In the US, hybrid corn replaced traditional openly bred corn varieties completely by the late 1940s.[27] Hybrid corn, which Zvi Griliches (1964; p. 370) referred to as "one of the outstanding technological successes of the century," has also practically replaced traditional varieties in Europe and is spreading to developing countries. In China over 90% of the total corn growing area is filled by hybrid plants.[28] Other crops have also increasingly been hybridized. Within 4 years of the introduction of hybrid sorghum in 1957 practically all US sorghum growers had switched to hybrid seeds.[29] Rice may be facing a similar fate. After Chinese researchers discovered how to hybridize rice in 1970, over 50% of the world market leader's[30] total rice area was soon covered by hybrid rice.[31] In Europe over 70% of all vegetable varieties are hybrids.[32] This development had the dual effect of marginalising farmers' role in seed-breeding in the affected crops and allowing seed corporations to "retain the exclusive right to reproduce and multiply a plant" (GRAIN 2000; p.19).

While advanced varieties were also increasingly bred for self-pollinating crops such as wheat, soybeans, and cotton, these could however still be saved by farmers and used for experimentation purposes. Especially in developing countries this practice is still common and in collaboration with other farmers new seed varieties are still being tested and occasionally farmer seed innovation still occurs.[33] Biggs and Clay (1981; pp. 322 and 326) succinctly summarize that "farmers will innovate within the limits of their technical capacity [and the access to materials such as their local gene pool] to adapt a technology to their productive circumstances."

The know-how and resource deficits, especially in terms of access to advanced germplasm, of farmers however function as significant hurdles to farmer-breeding. Take the example of Iraq. Before the recent US invasion, according to a 2003 FAO report, around 95% of the used seeds were saved by farmers. The vast majority of these seeds however were of poor quality due to the use of inferior cultivation instruments.[34] The development of more sophisticated breeding techniques such as tissue control beginning in the early 1950s, allowing scientists to breed even more advanced varieties, meant that seed-breeding had largely left the fields and moved into scientific laboratories.[35] This led to a gradual separation between plant-breeders, who developed and diffused new varieties, and the users of seeds, i.e. the farmers who planted the seeds. Seed companies gradually grew into large enterprises that integrated plant breeding (i.e. R&D), production, conditioning, and marketing functions.[36] Especially the large chemical companies—Monsanto, Bayer, DuPont, Shell Oil, Upjohn, Cargill and Ciba-Geigy—have become the dominating industrial forces. In the US for example, the four largest companies for each crop, accounted for 69% of corn seed sales in 1997, 47% of soybean sales and 92% for cotton.[37] According to the USDA Monsanto currently controls around 80% of the US soybean and cotton markets.[38]

While seed-innovation used to simply require an understanding of biology and pedology, the study of soil in its natural environment, the skill set and the instruments required to breed (e.g. absolute sterility) became increasingly sophisticated.[39] Biotechnology, with its genomics and genetic engineering, turned seed-innovation into a high-technology undertaking.[40] The former enables the identification of desired genes, to determine their function and to understand their interaction with other genes chemically as opposed to inspecting the grown plant. This method is much more reliable and faster. Biotechnology allows the manipulation of an organism's DNA by introducing or eliminating a specific gene. In this manner genes can be introduced from anywhere, not just from the same species, giving breeders a vast amount of possibilities that farmers do not have.[41] Biotechnology has further enabled the creation of vitamin and mineral enhanced crops, ones which contain less toxic elements and which are resistant to insects.[42] With traditional breeding methods and the limited resources (time, tools, money) available to farmers these innovations could not have been made.[43] Given that still around 40% of the world's crop, especially in Africa and the Middle East, are destroyed by parasitic plants and insects, biotechnologically engineered seeds may be the only economically feasible solution.[44]

The superiority of professionally bred varieties is illustrated by the Green Revolution. As farmers themselves seemed unable to increase global food supplies by the amount necessary to match the world's massive population growth, and thus to avoid major famines, high-yielding varieties (HYVs), primarily in rice and wheat, developed by various international research centers[45] in collaboration with private companies, were released and were adopted at record speed around the world (except in Africa). These HYVs dramatically reconfigured the shape of global agriculture.[46] While the high yield increases allowed massive famines to be avoided,[47] by making farmers dependent on HYVs, they also further marginalised the century-long practice of using and cultivating farm-saved seeds.

The vastly higher yields[48] and the quality enhancements[49] of professionally bred varieties would have been impossible via ordinary traditional breeding methods.[50] Consequently, the majority of Western fields and a considerable amount of developing lands were soon planted with professionally bred varieties. The case of high yielding rice that the International Rice Research Institute (IRRI) first introduced in 1966 via the IR 8 illustrates this. By the early 1990s average rice yields had increased by over 80% and "the new strains had [largely] replaced traditional varieties in much of Asia" (Douthwaite 2002; p. 164). While in 1970, about 30% of the rice and 20% of the wheat areas in developing countries were planted with modern varieties, by 1990, these shares increased to 74% and 70% respectively. While in Asia as a whole around 80% of the total rice area is planted under modern varieties, in some countries such as Japan and South Korea around 100% is covered by modern varieties.[51] In China modern wheat and corn varieties cover 70% and 90% respectively of the total crop area.[52]

Douthwaite (2002; p. 184) therefore concluded that "the Green Revolution has taken away the responsibility for plant breeding from farmers themselves and given it to breeders in public research institutes and private companies." Vernooy (2003) also characterized the situation as a "top-down system of agricultural research, where farmers are seen merely as recipients of research rather than as participants in it." While the Green Revolution's "induced innovation" (Ruthenberg 1985; p. 4) not only occurred without any user-innovation itself, it further led to farmers around the world increasingly becoming "passive consumers of germplasm" (Hope 2004) who plant the "finished" seeds, rather than being active innovators (Bellon and Morris 2002; p. 9). International research agencies for example frequently supplied farmers in developing countries with "biological packages" of high-yielding seeds (together with guidelines, tools and fertilizers) that were "not supposed to be modified" (Chambers et al. 1989; p. 46). Overall, modern seed innovation can thus be characterized as seen in Table 12.

The rise of genetically-manipulated crops will further reinforce this trend away from farmer-seed-innovation. In 2004 over 85% of all US soybeans, 76% of all corn and 45% of all cotton were already genetically manipulated.[53] According to some estimates, GM crops already account

*Table 12* Modern Seed-Innovation

| *Activity* | *Farmer* | *Supplier* |
|---|---|---|
| 1. Selection of source germplasm | | |
| 2. Trait development | | |
| 3. Cultivar development | | |
| 4. Varietal evaluation | | |
| 5. Seed Diffusion | | |

Source: Own Table

= Party engages in designated activity

for around 25% of the world seed market.[54] This trend is likely to continue as the strong annual growth rates of the global area of biotech crops of between 10–20% over the last years indicate.[55] Growth was especially strong in developed countries, arguably due to the lower bargaining power of farmers in such countries and the lower aversion to GM crops. Despite the currently existing resistance GM technology is here to stay,[56] not only because according to WTO-rules GM crops are freely exportable but also because it seems to be the only way in which the growing world population can be adequately fed. Additionally, they are strongly being pushed by plant breeding lobbyists, which has already lead to the lifting of the EU GM ban in April 2004 and to a gradual spread of GM crops, especially in eastern European countries such as Poland.[57]

## LEGAL CONSIDERATIONS

A further crucial barrier to farmer seed-innovation is the substantial strengthening of legal protection for seeds in recent decades. While in 1981 patenting and "other ways of monopolizing rents from innovation" were still insignificant in the agricultural field, this has since changed significantly (Biggs and Clay 1981; p. 321). Intellectual property protection for plant varieties began with the founding of the Union for the Protection of New Plant Varieties[58] (UPOV) in 1961. Gradually through the following years plant variety protection (PVPs) were introduced globally and patent protection was extended to all types of seed varieties around the world.[59] PVPs are a new form of intellectual property for seed-breeders that have been "described as patent rights with some missing attributes" (Cullet 2004). While PVPs also provide breeders with an exclusive right to produce, reproduce and commercialise protected varieties, including their export and import, they have traditionally contained much broader exemptions than patents. PVPs for example do not extend to private, non-commercial or experimental usages of seeds.[60]

Additionally, the so-called farmer's privileges[61] meant that farmers were entitled to save seeds to use in subsequent harvests.

Recently however, partly due to further corporate lobbying, the duration of PVPs was increased from 15 to 20 years,[62] PVPs were extended to all species, and farmer's privileges were curtailed in the EU and US.[63] While Articl 2 of UPOV 1978 stated that member states can only offer IP protection for plant varieties through PVPs or patents, under UPOV 1991 this restriction was eliminated to allow double protection. While in many countries biotechnological inventions were not patentable, US pressure has led many nations, such as Peru, Bolivia, Columbia, Ecuador and Venezuela, to change their patent laws.[64]

Plant patents, always dependent on the requisite research exemption of the national patent law at hand,[65] can be used to prevent most kinds of further commercial breeding using the protected materials.[66] While PVPs allow breeders to use a protected variety for the purpose of breeding and commercializing other varieties,[67] the original rights holder can prevent the commercialization of seeds that are "essentially derived" from his variety. The combination of patent and PVP protection can enable seed-developing companies to extensively protect their innovations.[68]

Furthermore, many laws now, directly or indirectly,[69] restrict farmers' rights to control and use farm-saved seeds to develop new varieties. EU farmers for instance, under seed-saving exemptions, have to pay royalties to PVP holders when they use farm-saved proprietary seeds.[70] Nearly all vegetables except potatoes are not included in this exemption and can therefore not be saved.[71] US farmers similarly may not sell, save, or breed proprietary seeds for commercial purposes without breeder permission.[72] Due to Western pressure various other countries were forced to adopt laws that limit or even prohibit the saving of proprietary seeds: most recently Sri Lanka, Cambodia and Iraq.[73]

Farmers are generally unable to meet the criteria for PVP or patent protection such as novelty, distinctiveness, uniformity and stability.[74] This can be illustrated using the case of Iraq, the "birthplace of farming" (Diamond 1998). Smith (2005c) clearly explained that "it is impossible for the seeds developed by the people of Iraq to meet these criteria." The seeds that Iraqi farmers employ are not new as they are the product of thousands of years of development. They are not distinct as the free exchange of seeds which has been practiced for centuries has ensured that specific characteristics are shared across local varieties. Such seeds are furthermore neither uniform nor stable due to the very nature of their biodiversity. Their constant cross-pollination with other nearby varieties ensures that local varieties are constantly changing and always adapting.[75]

In the EU this is very similar, as many farmers are neither able to meet the criteria for obtaining PVP protection nor the similar ones for registration on the EU Common Catalogue of Varieties[76] or of the National Catalogues. Without such registration seeds cannot legally be commercialized.[77] To be registered varieties have to be "clearly distinguishable" (Art. 5.1) and to "offer a clear [qualitative or technological] improvement" (Art. 5.4) from any other variety known in the Community, all of the seeds from the same batch have to be identical (Art. 5.2) and remain the same after successive generations (Art. 5.3), and finally possess satisfactory value for cultivation. According to the German *Bundessortenamt,* the German seed variety office, the mere evaluation whether such value exists takes 2 to 4 years.[78] Dr. Reinhard Müller, in charge of grains and cereals at the German Association of Plant Breeders (BDP), stated that these are "very tough requirements" to overcome.[79] Due to the lack of time, know-how and finances, Kästler (2005) has argued that it is practically "impossible to fulfil the criteria" required to commercialize seeds for farmers.[80]

Getting patent protection is similarly illusory for most farmers. The required inventive step often poses an insurmountable barrier. Consequently, multinational corporations largely control IPRs and R&D in this realm.[81] For example, over 80% of all field trials throughout the late 1990s were conducted by only four firms, with Monsanto alone in some years accounting for a third of all trials.[82]

As an increased amount of biotechnology[83] and plant patents and PVPs are being issued, the number of separate rights needed to produce a new agricultural innovation has escalated (Diagram 10). From 2001 to 2005 the total amount of US and global PVPs that were in force increased respectively by 40% and 21%.[84] In a survey of 120 US clinical labs 25% responded that they had stopped to perform certain types of genetic tests due to patent restrictions or excessive costs.[85] The situation had become so bad that US public institutions had to found PIPRA—the Public Intellectual Property Resource for Agriculture—to ensure that public sector agricultural research institutions will be able to overcome these substantial IP "barriers to commercializing new staple and specialty crop varieties" (PIPRA 2005). The presidents and chancellors of 13 key US research universities further explained that "limited or conditional access to a wide range of patented technologies has been identified as a significant barrier to the applications of biotechnology in the development of new crops" (Atkinson et al. 2003; pp. 174–175). Brian Wright, Professor of Economics at Berkeley, adds that "no new companies are entering mainstream agricultural biotechnology because of overly broad patents on genes and research methods" (Davidson 2004).

A good illustration of this problem is the golden rice case. Vitamin-A enhanced Golden Rice was a revolutionary new product that promised huge health benefits to millions of poor children in developing countries, where vitamin-A deficiency causes 500,000 cases of blindness each year and is a contributing factor in over 2 million annual premature deaths.[86] As the rice however, which was first developed by scientists at the Swiss Federal Institute of Technology, contained DNA borrowed from bacteria and daffodils its delivery to the poor was burdened by 70 patents belonging to over 30 companies, and six restrictive material transfer agreements

(MTAs).[87] Without the substantial public pressure, the special social interests at hand and the partnership with what is now Syngenta, the researchers would most likely not have been able to overcome these barriers.[88] Syngenta provided substantial organizational help, in testing, navigating the patent thicket and dealing with regulatory hurdles.[89] Currently, the partnership has established a network of licensees in which various universities and plant breeding associations have the right to conduct research in this area. While all commercial rights must be passed on to Syngenta, for humanitarian purposes the seeds are freely distributable.[90] Hope (2004; p. 188), therefore acknowledges that

> even if open-source biotechnology projects could succeed in generating useful technologies, the high costs of regulatory compliance would make the community dependent on large corporations for production and distribution, with the associated danger of appropriation and the ex-ante disincentive to contribute that this implies.

While it may be sensible to relax seed-laws, especially in developing countries to allow more farmer innovation,[91] the current regime is unlikely to be relaxed in the near future as "powerful economic and legislative forces" will want to preserve the status quo of the multi-trillion dollar food industry—of which seeds are the "central factor" (Fowler and Mooney 1990; p. 118).

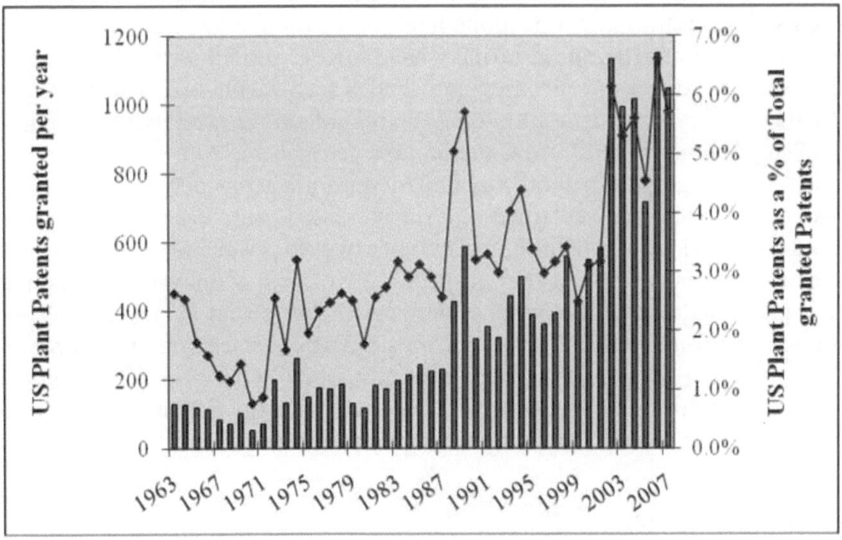


*Diagram 10* Annual US granted plant patents vs. total granted US patents. Source: USPTO 2008b

## MARKET BARRIERS

### Restrictive Contracts

Farmer seed-saving has caused a huge difference in the appropriability of R&D investments into hybrid and non-hybrid seeds.[92] To improve the appropriability of non-hybrid seeds, companies have used their market power to force customers to accept restrictive contracts. Monsanto, which has been referred to as the Microsoft of the plant world,[93] obliges any North American farmer who wants to employ its technology to accept licenses that allow such seeds to be employed for no other purpose than a single planting and harvesting:[94] Crop breeding, research or seed production[95] were expressly prohibited. Monsanto additionally forces growers to grant it a right to inspect their fields—for 3 years after purchase in case of Bt-cotton—and if farmers are found to saved and replanted seeds, they "are liable for a non-compliance penalty of 120 times the cost of the seed" (Barnett and Gibson 1999; p. 650).

Monsanto has successfully enforced these licenses against a wide variety of predominantly North-American farmers through intense investigations of farmers followed by attempted out-of-court settlements, and if all else fails litigation.[96] Monsanto alone had filed over 90 lawsuits against US farmers by 2005.[97] In the US case *Monsanto v Ralph*, for example, a farmer who had saved, replanted and resold seeds, was faced with a $3 million reasonable royalty jury award, that was upheld by a Federal Circuit Court.[98] One Tennessee farmer was even sentenced to 8 months in prison for growing 41 tons of GM cotton and soy in violation of Monsanto's patent and contractual rights.[99] Monsanto's actions alone have led to the bankruptcy of at least 9 farmers.[100] Several hundred farmers have further settled with Monsanto.[101] In regard to cotton such license finally ended seed-saving by US farmers[102] to further entrench the separation between manufacturers and users of seeds.

Percy Schmeiser (2005), a Canadian farmer who was forced by Monsanto to pay damages for patent violation despite allegedly never purposefully using GM seeds, has explained that the increased contractual ties of the big seed corporations and Monsanto's practice of even rewarding seed users who report other farmers who they suspect are violating Monsanto's rights, has created "a culture of fear" that is destroying the "social balance."[103] While in other regions such as Europe such restrictive contracts are far less prevalent[104] this may change with the increased spread of GM crops. Despite their debatable legality in other countries they still have the potential to discourage farmers from experimenting with proprietary seeds.

### Contract Farming

Economic changes have not only led to a decline in the bargaining power of farmers vis-à-vis their suppliers, but also vis-à-vis their customers. As crop processors face an increasingly high need to procure raw materials such as crops in the right quantity, quality and at the right cost they have exerted a substantial amount of pressure on both developed and developing farmers.[105] To ensure that the crops meet predetermined standards farmers have often contractually been obliged to utilize proprietary seeds as these led to more homogenous crops and unlike saved-seeds also come with quality assurances.[106] As German sugar refineries for example require a uniform crop they have contractually forced farmers to grow specified seeds and to spray particular chemicals.[107]

Such practice of contract farming, where agricultural products are produced under forward contracts between producers/suppliers and buyers, has become increasingly important in recent years, especially due to its promotion by major international donor agencies, multinational corporations and national governments.[108] While it is being pushed as it provides a suitable means to deliver inputs, technical information and finances to farmers, and thereby enables private investment into the crop growing markets, it is another reason why farmer-innovation has increasingly been sidelined.[109] Farmers who want to take advantage of the above benefits "must apply for a license and adhere to [specifically set] conditions" (Ruthenberg 1985; pp. 141–142). The typical contract is one in which the contractor supplies all the material inputs and technical advice required for cultivation, while the farmer supplies land and labor, leaving little or no space for experimentation.[110] Compliance is frequently monitored through field visits and post-harvest laboratory tests[111] and "non adherence leads to [draconic] sanctions like the withdrawal of credit, refusal to purchase or cancellation of the license" (Ruthenberg 1985; pp. 141–142).

In the US, contract farming today constitutes the primary method of handling major crops such as sugar beets, fruit, and "virtually all processed vegetables" (MacDonald et al. 2004; p. 9). In 2003 they governed 39% of the value of US agricultural production, a substantial increase from the 28% in 1991 or the 11% in 1969.[112] In an analysis of 420 production contracts in the US in the 1960s, 80% directly provided or controlled the "subject-matter input" and over 90% of the contracts involved some restrictions and/or monitoring of production practices. This led Harris and Massey (1968; p. 93) to conclude that contract farming "involves the transfer of a substantial proportion of the entrepreneurship function from the farmer to the contractor."

While there are various different types of farming contracts even in the least integrated/restrictive form, broadly referred to as marketing contracts, purchasers inform growers of the desired crop, variety, quality, timing, location and acceptable prices and thereby induce farmers to annually purchase proprietary seeds.[113] Da Silva (2005; p. 2) explains that the growth of contract farming is turning agriculture "into tightly aligned chains and networks, where the coordination of production, processing and distribution activities is closely managed." As competitive pressures have forced food processing corporations to improve the technical efficiency of producing, processing and distribution, the inputs, especially seeds, will continue to be closely controlled.[114] Contract farming is likely to continue to "marginalize the direct cultivators [i.e. farmers] who lose control over the production process" (Chandrasekhar and Ghosh 2003) and to essentially force the majority of farmers to "abandon traditional cultivation methods and products" (da Silva 2005; p. 18), to employ standardized farming methods.

## TECHNOLOGICAL BARRIERS

Despite of restrictive contracts and IP protection, the process of policing and monitoring farmers has proven to often be "inadequate, expensive to use [costing millions of dollars a year] and imperfect" (Srinivasan and Thirtle 2000; pp. 1152–1153). To remedy this seed companies are increasingly adding "marker genes" to their seeds' DNA.[115] Through sophisticated surveillance technologies, such as new DNA extraction methods,[116] proprietary seeds can be identified relatively easily. For example, John Deere and CropVerifeye developed a system called GreenStar, which enables farmers to electronically monitor and record nearly all aspects of their crop practices, including the genetic integrity and identity of seeds and crops grown under contract.[117]

As even with these increasingly sophisticated surveillance technologies costly legal procedures may be necessary to enforce contractual terms and IP rights, seed companies are investing into so-called genetic use restriction technologies (GURTs).[118] These can enable seed suppliers to program technological constraints into the genetic code of seeds. One controversial GURTs application is called "Technology Protection System" (TPS), dubbed "Terminator Technology" by opponents, which can genetically prevent seeds from germinating.[119] These "self-policing seeds" produce a crop for the farmer but no seed that can be saved (Burk 2004; p. 1558). In one variety of the technology, a "genetic switch" could even be introduced that will activate germination when the seed is exposed to a particular chemical and thereby could be used "to control the terms of seed usage from year to year" (Burk 2004; p. 1559).

The United Nations Environment Program (UNEP; 2003; § 7a and 15b) warned that "GURTS . . . might reduce basic knowledge and local innovation capacity of indigenous communities for crop improvement by reducing and limiting traditional seed exchange practices." Neth Dano, Executive Director of the Southeast Asian Regional Institute for Community Education, stated that

> this technology goes far beyond intellectual property. A patent expires after 20 years, but if Terminator seeds are commercialized it will give a handful of multinational Gene Giants a monopoly with no expiration date. This technology aims to eliminate the right of farmers to save seeds and do local plant breeding. (RAFI 1999; p.2)

While in theory GURTs could be circumvented in practice this is very unlikely as currently no "community of biological hackers [exists] who might either personally have the skill to circumvent biological lock-out coding or supply users with the tools to circumvent such code," and as such hacking is likely to result in direct and contributory patent infringement[120] (Burk 2004; p. 1571).

Due to the public outcry there is currently a moratorium on the commercialization of Terminator technology,[121] which however, a strong corporate[122] and governmental[123] lobby is trying to overturn. To put GURTs into perspective, they should be seen as "an induced response to the inadequacies and weaknesses of existing intellectual property rights institutions" (Srinivasan and Thirtle 2000; p. 1147), which are unlikely to vanish in the near future. Especially the t-gurt variety, where individual traits can be switched on or off, and the possibility to control the seeds' geographic and seasonal use seem promising.[124]

## SEED-INNOVATION IN GERMANY

To verify the above assertions we investigated the German seed industry. In-line with the trend in developed countries, the German farming industry has shrunken considerably in terms of the number of farms,[125] while the production value has remained relatively constant.[126] This indicates that only the most professionally operated farms are likely to survive. To determine the role that German farms played in seed-innovation we conducted various interviews with farmers, mainly representatives of regional farmer associations.[127]

While we found that many farmers still save seeds,[128] experimentation with such seeds was rare. Saved seeds are instead generally sent to a third party who reprocesses the seeds by cleaning these thoroughly to allow farmer to grow these in the next season.[129] In our interviews representatives of both German farmer and seed-breeding associations indicated that practically 100% of breeding is undertaken by professional seed-breeders. Leonard Keller, President of a regional Bavarian Farming Association of 5,000–6,000 seed-farmers, explained that he had not heard of a single example of farmer seed-breeding in Germany in the last 20 years. He emphasized that farmers merely plant seeds which are bred by professional breeders.

Ms. Newe of the Schleswig-Holstein Farmer Association stated that breeding has become "too complicated" and that "breeding techniques have become too sophisticated" for farmers who lacked the necessary know-how, skills and instruments. Dr. Carola Bühler, who advises the state government of Hamburg on agricultural issues, and Wilhelm Grimm, President of the Farmer Association of Hamburg stated repeatedly that it was "impossible" for farmers to breed/improve seed germplasm nowadays, unless this was in cooperation with universities or professional seed-breeders. Dr. Bühler referred extensively to the gap in farmer education in regard to seed-breeding. While nearly all farmers in Germany and other European countries have some type of vocational education and qualification, relatively few have university degrees.[130] As seed-breeding has become increasingly technologically complex and farmer education has not been concerned with seed-breeding it can be argued that the requisite educational systems have further reinforced the professional separation between farming and seed-breeding. Especially the necessary financial investments needed to breed advanced seeds today[131] were considered to be a high barrier by all respondents. Leonard Keller explained that purchasing seeds was more economical for farmers than breeding, especially due to the length of time it takes to develop a new seed variety: 10–15 years.

Representatives believed that the existing separation of labor between seed-growers and seed-breeders was 'absolutely sensible'[132] and that the vast majority of farmers were content with this status quo.[133] Dr. Bühler stated that "the job of a farmer is to grow seeds, not to breed seeds." This leads us to hypothesize that a main reason why so little or no farmer-innovation occurs in Western countries is that the due to the homogenous conditions in these countries plant breeders can develop varieties that fully satisfy farmers' needs.

## Organic Farming

In our search to find examples of farmer seed-innovation in Germany in the last decades, the only area we found in which farmer breeding still occasionally occurs was the realm of organic farming. This refers to a sustainable form of natural farming where chemical pesticides, synthetic fertilizers or genetically modified crops are not employed.[134] While according to Eurostat figures organic farming represented only around 4% of the total utilised agricultural area in 2002 in both Germany and the EU as a whole,[135] considerable growth is being predicted.[136] In the US, despite a

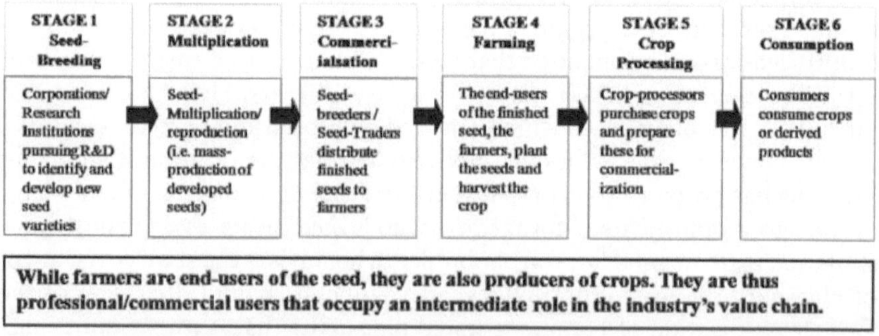


*Diagram 11* Modern agribusiness value-chain.
Source: Based on personal interviews, Asokan and Singh (2003) and Kesan (2000

substantial surge in recent years, organic farming still represents less than 0.3% of the total agriculture area. In most Asian and African countries it plays an even smaller role.[137] While the vast majority of organic farmers use modern varieties,[138] various hobbyists exist in this realm, a few of which even possess small research laboratories.[139] Occasionally these farmer-breeders through experimenting with saved seeds even develop improved organic varieties.[140]

Why do organic farmers still occasionally breed their own seeds? First of all, unlike regular farmers, they still have a need to do so, as private plant breeders currently offer only very few organic seeds.[141] Secondly, organic breeding relies "less on capital and access to technology" but on human labor and know-how (Ikerd 2000; p. 2). Nevertheless, organic farmers are still generally obliged to cooperate with organic research institutions/seed-breeders if they want to breed organic seeds.[142]

One noteworthy example is Karsten Ellenberg, who undertakes small potato breeding activities on his farm in Northern Germany.[143] In a small laboratory he experiments with old varieties from Peru, England and Germany. He calls this "conservational breeding,"[144] where the emphasis is on breeding and up keeping old varieties to sell the produce. While Ellenberg successfully sells red, brown, purple or even white potatoes[145] he does not commercialize the seeds, due to the legal barriers involved. Although in the 8 years of his breeding he has developed two new varieties, "Emma" and "Olivia," he has not applied for PVP protection for either of these. This is because Emma has too short of a dormancy period and as Olivia's' taste

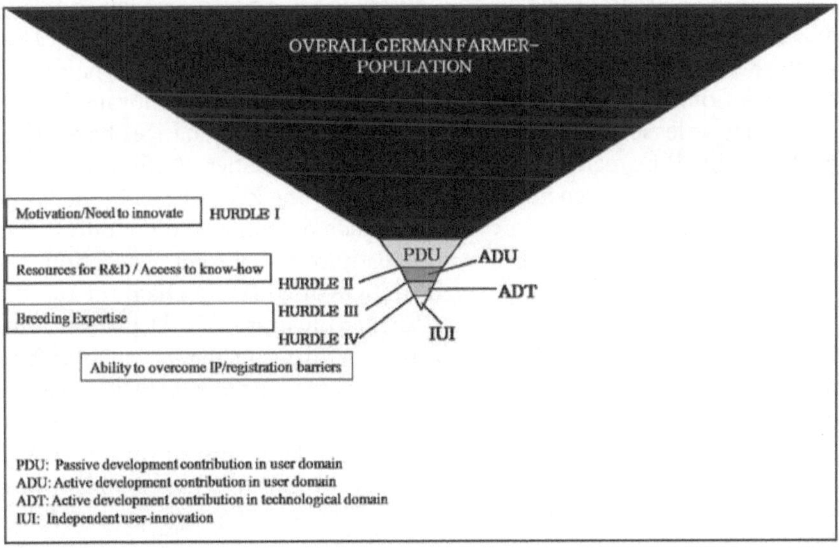


*Diagram 12* Swell model for modern plant breeding.
Source: Own diagram

and yield still has to be improved. Ellenberg, who wants to have a registered and protected variety on the market within the next 4 years, acknowledges that seed breeding has become complex and expensive. Nevertheless he asserts that in some cases organic farmers can breed new varieties if they are willing to invest the necessary time and resources. He states that the main problem is that "most farmers have lost the know-how to breed seeds," which he blames on their "dependence on plant-breeders."[146] Mrs. Newe therefore referred to his activities as "absolutely marginal."

While small scale organic seed production has been going for a considerable time in various European countries[147] this does not refer to actual seed innovation as unlike the resulting crops the seeds themselves cannot be commercialized. Ellenberg explains that the "legal situation at the Bundessortenamt [the German Variety Registration Office] is clearly inhibitory . . . [especially] the value examination is anachronistic and results in a patronizing of farmers and consumers"[148] in favor of professional plant breeders. To summarize, "most organic farmers realise that producing good quality seeds is a skill in itself, and they rather rely on professional seed producers" (Lammerts van Bueren 2002; pp. 78–79).

## FARMERS IN THE WESTERN VS. FARMERS IN THE DEVELOPING WORLD

Technological, legal and economic factors have dramatically transformed the seed-breeding industry over the last hundred years from being a pure user-dominated realm to being an industry in which farmers play at most a marginal role as seed-innovators. A comparison of the swell models in Diagrams 8 and 9 and the one above in Diagram 13 illustrates the dynamics of this evolution. While previously most farmers were able to innovate, it illustrates that at least in Europe and the US the discussed changes have effectively "lock[ed] farmers out of the seed system" (Kästler 2005; p. 10). Very few farmers seem to even have the need to enter the PDU phase. In the Western world, the very few organic farmers which do manage to not only create a new variety, but also to diffuse it, constitute the absolute exception—as indicated by the miniscule checkered triangle titled IUI in Diagram 13.

The statistics on genetic diversity further illustrate this decline. According to the UN's Food and Agriculture Organization the global switch to hybrids and HYVs caused approximately 75% of the genetic diversity of crop plants to vanish throughout the last century.[149] A report entitled *Nature's Cornucopia* found that in the US over 80% of seed varieties sold a century ago are no longer available.[150] After World War II indigenous wheat varieties in Greece, Turkey, Nepal, Spain, Pakistan, Eastern Europe, Libya and Iran have "virtually disappear[ed]" (Fowler and Mooney 1990; pp. xii and 68). China has lost nearly 90% of its traditional wheat varieties over the last 60 years.[151] Fowler and Mooney (1990) in their book *Shattering: Food, Politics and the Loss of genetic diversity* also indicate that for 75 types of vegetables approximately 97% of the varieties given on a 1903 USDA list of the then existing varieties were now extinct. Finally, while Filipino farmers once grew thousands of kinds of rice, by the mid 1980s only two varieties accounted for 98% of the area sown.[152]

As Hilkka Pietila, a Finnish politician, succinctly summarizes: "The last century of scientific breeding has divorced farmers from their own seed and effectively reduced the number of breeders from millions of farmers to a few hundred companies" (Fowler and Mooney 1990; p. 131). As a clear distinction between seed-manufacturers and farmers has formed we conclude that innovation in the seed-breeding industry is certainly not being democratized, au contraire.

Nevertheless, despite of the existence of strong barriers to innovation, farmer seed-breeding to tailor varieties to local conditions still exists to some limited extent in the developing world.[153] Various factors distinguish farmers in developing countries from western ones. First of all, finished cultivars make sense when a crop is grown in a "large, ecologically homogenous production environment" (Bellon and Morris 2002; p. 9), but are often unsuited for heterogeneous regions. While Western farmers are faced with relatively uniform farming land[154] and thus are satisfied by the seeds that are available from seed breeders, the huge heterogeneity in developing countries (in terms of soils and climatic characteristics), especially in Africa,[155] frequently means that such farmers have unmet needs. Both public and private seed-breeders tend to create varieties which, while useful for standardized conditions, are often not suited for very heterogeneous environments. This explains the bigger desire to experiment and tailor seeds to their land. Due to the sheer amount of farmers with unmet needs in idiosyncratic conditions, some may discover or cultivate varieties that are also useful for other farmers in their area.

The legal difference is also significant. Despite of the existence of TRIPS, the expense of obtaining global patent protection often means that companies refrain from patenting in developing markets, leaving local farmers free to develop their crops for commercialization in their home markets.[156] Although for most of the important crops that matter for food security in poor countries farmers' "freedom to operate" is not impeded, in regard to export-oriented cash crops, such as coffee, cocoa or bananas, this is a big problem.[157] Due to weaknesses in terms of legal enforcement, contracts that prohibit the saving of seeds are also less frequent in developing countries.

The biggest problem for developing country farmers is the lacking scientific skills and resources to innovate.[158] The high illiteracy rate for example poses a considerable constraint to farmer's improving their know-how and transferring these to their colleagues.[159] Nevertheless, the help and support of public research institutions, developmental aid and NGOs has helped farmers to participate in the breeding process.[160] As an example the International Maize and Wheat Improvement Center (CIMMYT) deliberately bred composite maize (instead of hybridizing it) to allow such farmers to save their own seed and adapt these to their local conditions.[161]

## HYPOTHESES ON THE DYNAMICS OF USER-INNOVATION

### Influencing Factors

On the basis of the previous discussion we will now present some hypotheses about the dynamics of user-innovation. We intended to demonstrate that the democratization of user-innovation argument is currently still somewhat incomplete. To start at the basics, we believe that a world without manufacturers is just as implausible as a world without users. While some factors have made user-innovation in regard to information goods easier, various factors have had the opposite effect. There will be a constant struggle between encouraging and discouraging effects which will allow user-innovation in some areas at some occasions to flourish, while at others to vanish.

There are two principle caveats to our findings. First of all, this study is not intended to be exhaustive. Other factors that we did not discuss may in some situations also influence user-innovation. Secondly, we need to point out that the limited amount of data available on user-innovation over time in various industries and the non-existence, except to some extent as obtained by our CDENT study, of data on the barriers to user-innovation, and their evolution over time, makes generalizations difficult. While further research is obviously required to verify our assertions, it may be cumbersome to gather suitable data to test our hypotheses.

> **Hypothesis 1:** *The degree of user-innovation changes over time as the barriers that users have to overcome before being able to innovate evolve over the years.*

A main part of our argument is that in a world that is characterized by increasing specialization in realms as distinct as journalism, law, science and the business world, it seems logical that we shall see an increasing separation between R&D firms and manufacturing firms. Just as the best inventors are not necessarily the best producers, those who can use or employ a product in the best manner, will often not be the best at improving it, as the example of professional athletes indicates. According to Baumol (2004; p. 195) specialization should ensure that "inventive activity will be undertaken primarily by the more effective inventor, while production of the resulting products will be undertaken predominantly by the more efficient producer."

In practice we are therefore likely to see a continuum of user-innovation. In some areas during certain points in time users will be the more efficient inventors while in other areas, or at other time, manufacturers will have an efficiency advantage. There is a range of continuums from 100% user-innovations in some areas—such as "ultraviolet spectrophotometry" in the early 1970s[162]—to 100% manufacturer-dominated innovation in other areas—such as in engineering polymers[163] or chemical additives.[164] Pavitt's (1984) analysis of 2,000 significant innovations in Britain since 1945 further demonstrates the huge differences in user-innovation in different industries: While across his sample around 11.2% of innovations came from users (he only looked at corporate users), in some sectors such as leather and footwear, motor vehicles and chemicals, the extent of user-innovation varied from 0–2%, in other fields such as in mechanical engineering and instrument engineering the figures were 15.3% and 19.3% respectively.

Shah and Tripsas (2004) mention four variables that will influence whether or not, and if so how, users decide to commercialize their developments: The user's estimate of the profitability of entering the product market on his own, the user's profit threshold, the manufacturer's estimate of the profitability of entering the product market, and the manufacturer's profit threshold. While the intellectual property regime and the newness of the market are critical in this respect, a wide variety of other factors such as the value of the improvement to other parties, the complexity of the product and the users' developmental, marketing or manufacturing skills are also likely to influence the decision making process. While Shah and Tripsas' (2004) discussion applies to the diffusion stage, numerous factors seem to previously affect whether or not a user can modify the technology/product in a useful manner or not. The manufacturers' ability to stifle and influence user re-invention is an especially important component which has largely been left out of the discussion.

The related field of innovation diffusion, which has interestingly hardly been considered by the user-innovation literature, has studied why, how and when an innovation is adopted by a group of users.[165] It is a process that involves social change and adoption therefore differs substantially for different innovations.[166] The diffusion literature has identified a few factors that influence the degree of re-innovation.[167]

- The nature of the social system and the compatibility with the current values and system.
- The type of innovation involved
  - Its complexity: "The complexity of an innovation, as perceived by members of a social system, is negatively related to its rate of adoption" (Rogers 2003; p. 257).
  - Its trailability, i.e. the degree to which an innovation can be experimented with on a limited basis.
  - Observability, the degree to which the results of the innovation are visible to others.
- The extent of manufacturers' efforts in diffusing the innovation.
- The resources of the adopters.
- The modularity of the innovation. Non-modular will be difficult to re-invent as the whole product has to be modified and therefore adopters cannot mix and match.

Holt (1988) further mentioned that the extent to which users, from a corporate perspective, should be integrated into the different level of the innovation cycle depends on five sets of factors:

- company factors—corporate culture/management philosophy, technology strategy, size, resources, internal reaction
- individual factors—attitude, education, experience of management
- product factors—importance of the product, novelty, degree of customization, place in the product life cycle
- project factors—expected cost/benefit of integrating users
- external factors—benchmarking, type of market, competence of users, access to users, distance to users, user stakeholders.

Jointly with the lessons learned from our discussion we can take these factors to summarize the principal determinants of the dynamics of user-innovation:

We identify 5 domain headings as being of the most important factors influencing the prevalence of user-innovation: Environmental factors in regard to where the activity is being undertaken, the specific technology involved, the market on which diffusion will occur, supplier-related and user-specific factors. As all of these factors, which we will briefly discuss, evolve over time and influence each other in this process, their interaction shapes the dynamics of the user-innovation environment.

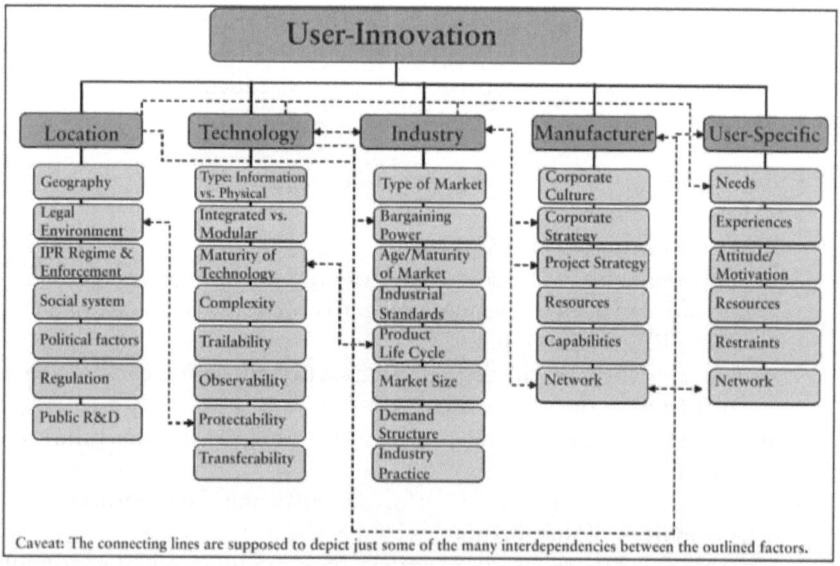


*Diagram 13* Non-exhaustive list of factors influencing the dynamics of user-innovation. Source: Own Diagram

## Importance of Location

In Nelson's (1993) edited work *National Innovation Systems* the huge differences between the different national innovation systems are discussed in-depth. Especially the substantial variations in R&D policies, industrial mix, educational systems, and the role of government, even between developed countries, were highlighted. Love and Roper (2001) also found significant national differences in the efficiency of R&D and innovation outputs. Various regions, such as in Silicon Valley in the US, the area around Cambridge in England, or Munich and its surroundings in Germany, have been known to yield superior innovative output.[168] While relatively little is still known about the forces that led to such uneven distribution of innovative activity and the creation of innovative hubs, it is known that the availability of different types of knowledge in different regions plays a crucial role.[169]

Knowledge spillovers, which often are important stimulants for innovation,[170] are generally, especially in regard to tacit knowledge, geographically concentrated.[171] Therefore, as the extent of know-how spillovers differs in other regions, different players will also have distinct types of know-how at their disposal. Local university R&D has for instance been shown to have an important impact on private sector innovation in the US.[172] Interestingly, Fritsch and Slavtchev (2005) found a considerably smaller effect of university knowledge on innovative output in their study of regional innovation in Western Germany. Therefore the effectiveness of university research also seems to vary across regions, time and most likely across industries. Krugman (1991) further indicates that US manufacturing and service industries are for instance highly localized and are therefore able to benefit from strong economies of scale, low costs of transportation, qualified employees and technological know-how spillovers in general. He adds that "concentration of industry [in a specific location], once established, tend to be self-sustaining" (ibid. p. 98). Access to research institutions, laboratories, finances, qualified employees, product demand in a region were found to be critical reasons for the establishment of innovation clusters in the Spanish and German medical devices industries.[173]

As Lüthje et al. (2005) found that user-innovators almost always utilize "local" information, i.e. know-how already in their possession, in their innovative endeavors, the different availability of know-how in different regions will in all likelihood mean that users in different areas have different abilities to innovate. This is especially the case as free-revealing will often be regionally constrained.[174] This can for instance be due to strong regional complementary assets or as in the case of US patent donations, which led to tax benefits,[175] due to the prevailing legal/political regime. In respect to invention by individual inventors, Jewkes et al. (1969) argued that in America the environment (better public research institutions, better funding opportunities, etc) as well as the personal characteristics of the US inventors (much more likely to tinker), allegedly makes individual inventing in the US more likely than in other countries. We therefore hypothesize that user-innovation will also differ on the basis of national cultural differences. Gelijns and Rosenberg (1999) for example found huge differences in user-innovation for CTs, MRI and endoscopic instruments in Europe/US and Japan. While in the former user-innovation was strong in the early stages of the technologies, principally due to the their strong research universities, in the latter users played at most a marginal role as the large manufacturing giants with their strengths in consumer electronics, only commercialized the technology at a later stage

These studies seem to indicate not only that innovation differs across different regions, but also that different factors influence the extent of innovation differently in different locations. Consider differences in regulatory regimes that can impact upon medical device user-innovation. Furthermore recall the case of Blair Simmons who had to stop his work on cochlear implants due to social pressure. Promising results obtained by other researchers gradually changed the reluctance of the medical profession to encourage such research and by the end of the 1970s, cochlear implantation had even acquired “a degree of credibility in professional ontological circles” (Blume 1995; p. 116). This demonstrates how a social environment can change to become conducive to user-innovation. It seems that not only are early adapters and tinkerers, at least initially, considerably restrained by the social environment in which they operate, but to some extent they can also influence such environment. This dynamic nature of the interaction between the environment and its institutions is of critical importance in understanding the evolution of user-innovation.

The large differences between farmers in developed and developing countries, and the impact of the different IP regimes, legal enforcement possibilities and developmental support by research institutions provides further support for our assertion. In this realm a further obvious influential factor is nature: The huge differences in the soils and climatic conditions of different countries influence the functional sources of innovation. Different natural environments, especially in regard to their homo- and heterogeneity, have had a large influence on the need and ability (e.g. in terms of seed-saving) of users to innovate. Other examples of the geographic concentration of user-innovation include Lüthje et al.’s (2002) finding that the “North Shore” of the Americas was a hub for innovative mountain-bikers and the new group of hackers working on developing plug-in electric cars, which is mainly concentrated in California, allegedly due to the state’s focus on environmentally friendly emission-systems and “its love of electric cars” (The Economist 2006c).

> **Hypothesis 2:** *Location will have a major impact upon both the extent and type of user-innovation. As the constituents of a location change over time, so will the functional locus of innovation.*

It does have to be acknowledged that the internet with its virtual user communities has, especially in regard to software and information-goods, significantly affected the influence of location.[176] Nevertheless, as Dahlander (2006) indicates in respect of the free and open source software (FOSS) locality can remain an important consideration:

> User-innovation networks are not necessarily local, but a way of getting access to distributed resources. The need for face-to-face interactions is not to be neglected. In fact, my case studies showed that firms use local communities initially to tap into global networks. When initiating communities, local interaction is important in setting the first contact and establishing practices.

## THE INFLUENCE OF TECHNOLOGY

The characteristics of the concerned technology will have a tremendous influence on the ability of users to modify it. The way in which a product is designed "can influence how much control manufacturers exercise over innovations that flow from use by early and even later adopters" (Finkelstein et al. 1995; p. 136). The most obvious example is the difference between open- and closed-source software. In the CDENT case study we saw that while 40.9% of respondents believed that they would be able to improve the source-code if it were open-source, as the software was closed, no improvements to the software were made. Farmers similarly were unable to improve hybrid seeds. The technology will influence not only the need that users may have to modify it, but also the type of skills they need to amend it, and how far its creators can control subsequent use. In regard to various realms such as medical devices, seeds, sports equipment and wind turbines we saw that as a technology matures, the amount of user-innovation declined. This leads Douthwaite (2002; p. 128) to conclude that user-innovation is only feasible if the technology is not overly complex at the beginning to enable a sufficiently large amount of users to develop their skills and thereby help with the gradual improvement of the involved technology. Ogawa (1998) also found that at least in the Japanese convenience store industry, innovations requiring a rich understanding of needs were generally carried out by the users, while innovation that required a rich understanding of technology were often carried out by manufacturers. The development of the automobile also illustrates this. In its early days average users, such as rural farmers or middle-class American leisure travelers, were able to substantially improve cars and to develop additional applications. Rural farmers for example adapted the automobile to run laundry machines, butter churns, water pumps, wood saws and to develop improved vehicles, such as tractors, to suit their different transportational needs.[177]

Middle-class travelers developed a variety of innovative automotive accessories such as radiator hoods, safety devices, interior heaters, automobile tops, trunks, reclining seats and even electric ignition to their cars.[178] As cars became increasingly high-tech and computerized, the skills required to successfully tinker with these have also changed. The professionalization of the car tuning industry illustrates that modifying a modern car has not become easier over the years.[179] The average driver for example is generally no longer technologically able to improve his car. Finally, the majority of respondents to our CDENT study responded that it had become more difficult to amend the current version of CDENT than its predecessors.

> **Hypothesis 3:** *User-innovation will in many realms be inversely related to the maturity of the technology in question.*

A further crucial consideration is whether the product in question is integral, i.e. consists of a single component such as many drugs or seeds (or is multi-component but there is no clear mapping between a single component and its functionality), or whether the product is modular, i.e. made out of distinct building blocks with particular functions.[180] Modular products are much easier to improve for users than integral ones as with the latter "it is difficult to adopt [and modify] one element without adopting [and modifying] the other" (Rogers 2003; p. 186). CDENT's non-modular design contributed to the difficulty of user-re-innovation. Although modular products could allow users to modify individual units of the program to tailor them to their needs they are currently still relatively rare as these are "much more difficult to design than comparable interconnected systems" (Baldwin and Clarke 2004; p. 153).

Whether a product is modular or integral is also likely to affect the scope of its patent protection. The ability to protect an integral product by a single strong patent can make it very difficult to legally circumvent such protection as patents in the pharmaceutical industry illustrate.[181] Every minor modification of the product could constitute an infringement. While modular products generally allow individual components to be modified without infringing an overall patent, the ability to erect a patent fence by protecting multiple components can also make it difficult for user-innovators to legally modify.

Another important influencing factor is the ability to experiment with a technology. As user-innovation requires tinkering with a product or technology,[182] technological or practical barriers that burden experimentation are likely to have a substantial adverse affect on the ability to modify. This can be demonstrated by looking at hybrid seeds or at CDENT's closed source-code. In both cases technological barriers made it practically impossible for users to experiment with their respective products as a result of which user-innovation declined dramatically. In respect of hybrid seeds, if a public sector organization wins the R&D/patent race for apomixes, the ability of some plants to clone themselves, and introduces this into food crops, the status quo could be reversed again. In such scenario farmers could potentially replant hybrids and by experimenting with these potentially cultivate promising hybrid varieties.[183] Similarly, as the results of the CDENT study show, the closed-source code was considered to be a considerable barrier to user-tinkering.

> **Hypothesis 4:** *The less users can tinker with a technology, the less likely they are to improve it.*

A further important distinction is that between information goods and physical goods (von Hippel 2005a). While pure user-innovation is often relatively easy in regard to information goods such as software, it will be more difficult in regard to physical goods. This is principally because of the substantially higher investments that have to be made into R&D, prototyping, manufacturing and diffusion for physical goods. We argue that the democratization argument is currently more suited to information goods than to physical goods. Seeds for instance are physical goods in which currently user-innovation is marginal at most.

The observability and the transferability of a technology can also influence user-innovation. If users can observe and record the results of their tinkering and pass these on to third parties, such as other users, improving a technology is more likely to be feasible. The more tacit or non-transferable the results (cf. Ogawa 1998) the more difficult will be group user-innovation. While in regard to some technologies, such as chemicals, it is relatively easy to articulate the know-how embodied in the underlying technologies, in other areas such as electronics "information is highly context dependent and cannot be crystallized to abstract generalizations" (Anand and Khanna 2000; p. 127). If users could not easily diffuse the results of their experimenting with OSS, user-innovation would be very difficult. Such observability is especially problematic in respect of farmers in developing countries who often lack the ability to record their cultivations, not to mention the ability to recreate or transfer them (Rhoades 1989). Patents in many realms can also hamper the transferability of the results of user-tinkering.

## INDUSTRY DETERMINANTS

Industry specific factors are also important determinants of user-innovation. Voss (1985; p. 114) explained for instance that "the role of users varies, it seems, depending on specific conditions within each industry." Anand and Khanna (2000; p. 127) add that "the ability to appropriate gains from innovation differs systematically across industries," which according to von Hippel (1981) is likely to influence whether users or manufacturers will be the more suited innovators. Another common distinction of different types of industries that is likely to influence the ability of users to innovate is the one between "complex" and "discrete" product industries on the basis of whether a new product is comprised of numerous separately patentable elements or relatively few.[184] For example, while electronic or semiconductor products are typically comprised of a relatively large number of patentable elements and are thus classified as complex, new drugs are comprised of a relatively discrete number of patentable elements.[185] As discussed this can influence the number of IPRs that tinkering users wanting to diffuse their results may have to circumvent. The likelihood of being sued also differs substantially between different industries.[186]

The type of market, i.e. whether we are dealing with an industrial or a consumer market, is also likely to be an influencing factor. Pure hobbyists are likely to have a different ability, time and set of resources than professional users. Corporate users, such as the design firms in the semiconductor industries, are likely to have different prerequisites to innovate than a doctor or a farmer. Lüthje and Herstatt (2004; p. 560) summarized that "the type of benefit can differ between different markets and user-populations."

The size of the market and the heterogeneity of customers also play an important role as they influence the available rents. In a large market with many heterogeneous users, user-innovation is much more likely than in a small market with homogenous needs (von Hippel 2005a). Whenever manufacturers can benefit from economies of scale and/or scope by developing standard products they are more likely to be the dominating innovating force than users.[187]

The shape of the value-chain in an industry is also crucial. In some areas, such as semiconductors, users play an integrated role in the innovation generation process, while in other areas like farming, no such role currently exists for most users. The length of the product life-cycle also seems to matter. We propose that high R&D investments and long payback periods adversely affect user-innovation. In areas such as pharmaceuticals or seeds, where high investments have to be made into research and development, we found that user-innovation was low, while in areas such as software, where such investments where considerably lower, the rate of user-innovation was higher. One reason for the difference in seed- and livestock breeding in Germany, where farmers were still actively involved, was that the payback time was considerably shorter for livestock as the results of the crossing could immediately be seen in the next generation.[188]

Furthermore, the costs to manufacture and commercialize a technology are likely to be important. Software and information-goods can almost be diffused for free via the internet,[189] while the manufacturing and diffusion of improved hardware will generally be more cumbersome.

> **Hypothesis 5:** *The higher the R&D/manufacturing/commercialization investments that have to be made and the longer the payback periods, the lower user-innovation will be in an industry.*

A further important influencing factor is the bargaining power of users vis-a-vis manufacturers. While to some extent users and manufacturers will always be dependent on one another, the extent of this interaction will vary substantially between industries. While in many industrial goods sectors there is considerable evidence for close interaction, in many consumer goods sectors little innovation interaction has been found.[190] Some industries, such as seed agriculture, will be characterized by dominant manufacturers which can limit or eliminate user-innovation, whereas in others, such as the medical device industry which is mainly characterized by small and medium players,[191] the users may have a stronger bargaining power. The CDENT study indicated that even within industries the differences in terms of bargaining power can be considerable. As in the realm of computerized dentistry the manufacturer holds a very strong market position, its ability to control user-innovation in this realm was substantial.

This bargaining power can obviously change over time. The higher the competition in the market the less power manufacturers will have over users. Consider for instance the difference between US and European seed farmers. As there is currently more competition in the European markets, such seed companies have not yet been able to force farmers to accept restrictive seed-licenses. Evolvements in technology can further reshape the bargaining positions. Developments in ICTs, especially the internet, have for instance, by granting modern users much better access to important information such as product prices, quality and experiences of other users, increased the bargaining power of many types of users.[192]

Industrial standards and practices can also stifle or encourage user-innovation. The later refers to whether the dominating players in an industry generally collaborate with users or not. In some areas as we saw in regard to seed-suppliers the major players did not desire users to participate in the breeding process. In other areas such as medical devices it is much more common to collaborate with users. The CDENT study again indicates that such behavior is neither absolutely necessary nor universal. Although the Sony Aibo case or Mollick's (2004) parasitic-users demonstrate that user-innovation can occur even it is not desired, the industrial willingness to enable or burden user-tinkering can significantly affect such activity.

Abernathy and Utterback's 1975 model of product and process innovations illustrated that the ability of users to innovate is also likely to depend on the maturity of the market, with user-innovation being more likely in the early than the later phases. Hienerth (2006) and Baldwin et al. (2006) demonstrated this for the rodeo-kayak industry. Lotz (1992) also found that in the mature area of medical disposables users play a minor role as the fierce price competition forces manufacturers to focus on cost-efficiency and process innovations.

**Hypothesis 6:** *A high amount of market power of incumbent companies and their resulting bargaining power vis-à-vis users has the potential to decrease the amount of user-innovation.*

**Hypothesis 7:** *Restrictive existing industrial standards and practices can lower the amount of user-innovation.*

**Hypothesis 8:** *As an industry matures, especially if a dominant design evolves, user-innovation is likely to be adversely affected.*

## SUPPLIER-RELATED FACTORS

We affirmatively answered the question of Lüthje and Herstatt (2004; p. 562) of whether manufacturers can actively change the cost and benefit expectations of users in the innovation process. Manufacturers in many realms may not desire and therefore attempt to prevent user-innovation. Holt (1988) explained that the strategic question of whether or not to collaborate with users will depend on factors such as the corporate culture, corporate strategy, the strategy of the particular project at hand, and the resources of the company. While for IBM it made strategic sense to donate 500 of its US software patents to the open-source software community in 2005[193] as this supported its business model, enabled it to compete with Microsoft[194] and allowed it to receive US tax benefits, this has obviously not been the case for most patent holders.

The ability to limit or encourage users will also depend on the technology at hand, the firms' capabilities and the needs and desires of its partners (i.e. its network). CDENT's closed source code, Monsanto's contracts to prohibit seed-saving, GURTs, restrictive market standards, enforcing IPRs against programmers or farmers, or stigmatizing hackers are just a few examples of how suppliers can in certain circumstances raise both the actual and expected costs and benefits of user-tinkering. If as a reward for tinkering with a technology, the user can expect a lawsuit, he is likely to think twice about such behavior.

**Hypothesis 9:** *Suppliers can affect the cost/benefit of user-tinkering and therefore both stifle or support user-innovation.*

## USER-SPECIFIC FACTORS

Different corporations have been found to have substantially different abilities to translate new external technology into their operations.[195] Not just their ability to learn but also their ability to imitate third party developments varies dramatically.[196] Users will similarly differ substantially in their ability to learn from their environments, due to the substantial differences in terms of their educational backgrounds, experiences or resources.

As improving a technology or product requires a larger amount of know-how than mere absorption or use, whether users will be able to remain at the frontier of knowledge will depend on their investment of resources and continuous updating of their skills, know-how and instruments. Doctors who in the 1970s wished to remain at the frontier of knowledge regarding implants for example needed not only to understand biology, but also to increasingly have a grasp of engineering.[197] Today informatics, biotechnology and nanotechnology have also become very important in this realm.[198] How many practicing doctors will be able to innovate at the nanoscopic level, given that this requires an advanced set of laboratory equipment, remains an open question. The access to suitable networks will be crucial to overcome such technological complexity.

Nagel (1993) in his study of the electronic circuit board industry therefore found that lead-user status changes over time.[199] Few users have indeed been found to develop more than one major innovation.[200] Studies undertaken in the scientific instrument (von Hippel 1988—87.5%, Riggs and von Hippel 1994—71%), in the semiconductor process equipment (von Hippel 1988—67%) or in the sports equipment industries (Shah 2000—100%) provide empirical support for such allegation that the vast majority of users developed just one major innovation.[201] As the ability of a user to innovate therefore seems to depend on the right combination of environment (economically and legally), need, motivation, resources and ability, as time goes by different types of users are likely to be able and willing to innovate.

Furthermore, users with a strong internal R&D department or close links to academia are for example more likely to possess a high absorptive capacity.[202] Lettl et al. (2006) suggested that professional users such as medical doctors may have an advantage over hobbyist tinkerers, especially in terms of access to finance, complementary interdisciplinary know-how and technological capabilities. Although hobbyists may arguably have a more open mind, less politics to tie him down and potentially, in case of a student for example, more free time to tinker, the ability to network often seems essential to success. Consider the open-source movement and the ability to instantaneously and freely diffuse the results of user tinkering. Before the development of the internet networking seemed to be a major problem in this realm.[203] The limited networking of Farmers and the fact that dentists as CDENT users rarely operate in a hospital setting may have been one of the reasons why both frequently lacked the skills to overcome the barriers to user-innovation.

**Hypothesis 10:** *The degree of user networking in respect of the technology in question is likely to be positively correlated to the amount of user-innovation in that realm.*

The size of the user firm will also frequently be of considerable importance. Roy et al. (1968; pp. 100–101) in an in-depth study of the adoption of agricultural innovations among Indian farmers found that there was a strong positive correlation between the size of a farm and their employment of new high-technologies. An increased size of farms is even considered to be a "necessary precondition to modernization." While this simply refers to adoption, it is likely to equally apply to some extent to re-innovation. Archibugi et al. (1986; p. 7) in a survey of the diffusion of innovations in Italy, also found that "the share of innovating firms increases with firm size in every sector." Bigger firms generally, although there were substantial industrial differences, tended to develop more innovations in house than smaller ones. In comparison to companies with 20–49 employees, companies with over 500 employees tended for instance to generate over 50% more innovations internally.

> **Hypothesis 11***: The type of users of a product, with their different experiences, needs, resources and restraints will have a critical influence on the amount of user-innovation in that realm.*

# Part D

# The Paradigm of Licensing to Innovate

After having scrutinized the barriers to user-innovation and their dynamics we would now like to assess one potential effect of these evolvements. We made the argument that "the relationship between the consumer and his or her products is gradually changing" (Lockton 2005; p. 33). Users today have an inherently complex relationship with their products. From co-creating to customizing to testing to merely using products—users today are in some form involved in practically all stages of the innovation process. Our previous discussion, rather than providing support for von Hippel's democratization of user-innovation, indicates that a more multi-faceted and ambivalent characterization may be more appropriate in numerous realms. While in some areas user-innovation may have become easier, generally the amount of control, especially technologically, that could potentially be exerted over the use of a product has increased. Indeed, there are various indications that in some areas such as in the purchase of a car, the growing of seeds or the use of medical devices, we may be moving towards a scenario where consumers increasingly may no longer purchase products themselves, but rather purchase a set of clearly specified rights[1] and thereby "effectively license the functions those products provide" (Lockton 2005; pp. 33–34).

This part will build upon the previous discussions and commence by introducing the possibility of a permission culture, which is characterized by selective user-innovation. This is not intended to be a universally applicable model, but rather as a thought-provoking extension of the previous discussion that is likely to be relevant in certain fields. Subsequently, the goal is to identify an instrument that can be used to allow users to overcome the barriers to user-innovation. The argument will be made that the license may be such a tool.

# D.1 The Permission to Innovate

## THE PERMISSION CULTURE

According to Lessig the term "permission culture" is the opposite of a "free culture." "Free" in this sense means that one can use the idea, technology or know-how without the permission of anyone else (Lessig 2004). To be more specific, a permission culture is a scenario in which creators are allowed to create only with the permission of creators from the past. It is the opposite of the democratization of user-innovation, as it refers to selective rather than universal user-innovation.

The main reason why Lessig (2004) believes that we are increasingly moving towards such a culture is that it has become increasingly difficult for companies to rely upon the public domain. He states that "the architecture and law that surround the internet's design will increasingly produce an environment where all use of content requires permission" (Lessig 2004; p. 277). Although Lessig's argument sounds extreme there seems to be ample support for his view in the existing literature. Jane Ginsburg (2000), a renowned copyright scholar at Columbia Law School, explained that a fundamental shift is happening in which we will stop owning and start experiencing works as allowed and controlled by the entertainment industry. In *The Age of Access* Jeremy Rifkin (2000) proposes that while ownership of physical property is becoming increasingly less important, the access to networks and valuable information will become the key to economic and social success. He writes that while the trenches between those who own property and those who don't are sizeable, the differences in power and freedom between the networked and the non-networked are likely to be far bigger. Harvard Law School's Jonathan Zittrain (2000; p. 1218) argues that we are slowly moving from an "Era of Promiscuous Publication," in which copyright enforcement was doubtful due to new technologies enabling the instantaneous and almost costless wide-spread copying of software, music, videos, etc., to an "Era of Trusted Privication," where copyright may no longer be necessary due to tight technological controls: "Well-enforced technical rights architecture would enable the distribution of information to a large audience—publication—while simultaneously . . .

not releasing it freely into general circulation—privication." In other words under such a scenario, licenses[2] to use information for a specific purpose would become necessary. Zittrain (2005; pp. 18–19) explains that such a scenario has become practically inevitable as a) the openness of computers and the internet have led to "to unacceptable [security] vulnerabilities," b) as "the internet and PCs attached to it [that enabled much easier user-innovation] threaten creative destruction to settled interests," such as to intellectual property rights, and finally c) as many key market players are keen to become "gatekeepers for successful products rather than platforms for all comers." The European Commission Information Society (ECIS 2004; p. 2) as well as the German government[3] implicitly confirm such prediction: "The way forward is a system based on existing exclusive rights backed by technologies that ensure a secure environment where such rights can be licensed and enforced."

The evolvements in the realm of trusted computing support this trend and indicate that a permission culture can extend far beyond the boundaries of copyright protection. Through the combination of legal, economic and technological barriers, such systems can considerably constrain the way in which users can utilize both protected material and the underlying hardware. Such systems, by substantially driving up both the financial and temporal costs of reverse engineering, therefore have the potential to substantially increase the importance of obtaining an ex-ante permission to experiment, as Hal Varian and Pamela Samuelson, both distinguished professors at Berkeley, argue.[4]

The IPR thicket clearly plays a principal role in such permission to innovate culture. In this respect it should be noted that "virtually all of the basic inventions which transformed [the US] economy have been involved in patent controversies" and needed some form of licensing accord to be resolved (Greenleaf 1961; p. 88). A simple illustration of the power of such thicket is that even the US government is subject to it. In the mid 1990s, when it wanted to develop its own public key encryption technology, it realized that the main patents in this realm where owned by a partnership between RsA and Cylink called "Public Key Partners." Anyone who was involved in this realm in the US had to obtain a license to the dominating patent.[5] Out of fear of direct or contributory infringement through the taxpayers that would have used the resulting system, even the US federal government had to negotiate the permission to innovate.[6]

In Japan, the patent thicket is even worse, as "patent flooding," i.e. the practice of filing numerous patent applications around an existing patentee's core technology, is even more common.[7] Cohen et al. (2000; p.1349) who distinguished between complex and discrete industries depending on the number of separately patentable elements that were needed to commercialize a new product, found that "in Japan, it is conceivable that almost all products are 'complex' . . . because in Japan patents encompass fewer claims, and the claims themselves tend to be interpreted narrowly, generating more patents per product and, in turn, widespread technological interdependence." Consequently, innovation in most Japanese industries requires licensing.

To illustrate the importance of obtaining permission to innovate in today's time we have to acknowledge that there seems to be a positive relation between strong IPRs and the amount of licensing.[8] The stronger patent rights are, the more difficult it is to invent around or ignore them, and therefore the more important it will become to obtain licenses to experiment. Martinez et al. (2004) for example found that 48% of the European companies responding to their survey believed that the likelihood of research projects being blocked (not started or stopped) due to patents from competitors had increased in the last 10 years, while only 2% stated that it had decreased. This is especially important given the trend towards multi-technology products that are covered by multiple patents often held by different entities. In complex product industries such as automobiles, aircrafts, semiconductors, computer hardware or computer software, IPR bottlenecks have frequently made licensing between the initial and follow-up innovators necessary.[9]

To exploit such scenario firms often patent complements to preclude rivals (and users) from being able to commercially use certain technology, and to obtain the "freedom to operate" for themselves.[10] Edler et al. (2004; pp. 83–86) confirmed this with regard to the massive surge in patents in Germany throughout the 1990s, during which the number of German patent applications at the European Patent Office more than doubled: "The blocking of competitors by means of patenting . . . [is becoming increasingly important] as patents are used to maintain one's own room to maneuver and at the same time to limit that of the competitors." Cohen et el. (2002) found that 78% of US respondents in complex industries used such strategy of filing blocking patents to prevent other companies from innovating without seeking a license. In Japan 82% of respondents in both complex and discrete industries admitted that they pursued such strategy.

Further survey results confirm these findings. Cockburn and Henderson (2003) found that for 98% of the questioned US companies "maintaining freedom to operate [obviously including freedom to innovate] in core technologies is a key criterion in a successful patent strategy." Bessy and Brousseau's (2000; p. 68) study of 50 representative French licensing agreements also indicates that firms patent and cross-license to preserve their "freedom to innovate." The finding that the most important reasons for licensing in Bessy and Brousseau's survey were "to access a technology to immediately use it" and "to take an option on a technology to benefit from (future) room for maneuver" indicates the importance that licenses can play in the innovation process. Given that the extent to which the combination of market power, legal regulations and technological developments has allowed companies to exert more control over their intellectual assets, we argue that in many areas innovation could increasingly depend on licenses.

It is important to clarify that the permission culture refers not to a scenario without user-innovation, but on the contrary to a scenario where those companies who can best tailor the innovative abilities of users to their own strategy are likely to outperform.

## FREE REVEALING AND THE INFORMATION COMMONS

Free resources have always been central to innovation, creativity and democracy.[11] Free revealing is of crucial importance in the establishment of the information commons that is generally needed by user-innovators to experiment and tinker. Without such a commons from which users can obtain know-how and needed instruments the democratization of user-innovation is likely to remain an inspirational theory. It should be added that while free revealing is critical, by itself it is not sufficient to enable user-innovation. Winter and Szulanski (2001) for example demonstrated that the replication of well-known organizational routines at new sites was both difficult and expensive. Levin et al. (1987) provide some empirical support for the assertion that in most industries only a few firms are capable of duplicating new products and processes: On average only 3 to 5 firms were capable of duplicating a major process or product innovation.

Free revealing refers to the voluntary giving up of all intellectual property rights to a technology and thereby freely and equally making it available to all parties.[12] In other words free revealing turns a technology into a "public good" that everybody can use (von Hippel and Krogh 2006; p. 295). From a commercial perspective free revealing can have one of 2 main benefits: 1) to grow the user-base for a technology, thereby creating and developing the market for complementary goods and services[13] and potentially creating a de facto industry standard; 2) to increase the value of the innovation for the innovating party.[14]

The main reason why we believe that we may be moving towards a permission culture is that many needed resources are not freely available. According to Benkler (2002; pp. 95 and 98) "a larger ration of new information will be owned, further decreasing the availability of pertinent public domain materials" that users could use for their innovation attempts. While some form of a public open domain certainly exists, many scholars believe that "instead of being regarded as public information, technology might be more conceptualized as a quantum of knowledge retained by individual teams of specialized personnel" (Rosenberg and Frischtak 1985; p. vii).

While large amounts of information and know-how, from source code, to industrial news to financial data, is available for free on the internet, the most valuable content frequently is only available for subscribers who accept certain terms and conditions, usually including some form of payment.[15] Good examples are the large content databases such as Proquest and Ebscohost who, through public access restrictions, are even "enclosing" publicly funded research and only making it available to those who pay substantial amounts of money (Poynder 2003b). Cutting-edge scientific, industrial or financial know-how that is contained in various journals or databases can be so expensive that only professional users can afford them.[16] Melissa Barr (2003) explained that the 2 biggest providers, LexisNexis and Westlaw, have practically "cornered the market for legal research materials." By adding editorial comments to case data they have turned free information into a "paid-for proprietary product" that is protected by copyright. Through their "ownership of proprietary search and navigation tools" the large content owners can thereby "use copyright law in order to appropriate public-domain data and then enjoy a virtual monopoly on its distribution" (Poynder 2003a; p. 38). Legal developments, such as the European Databases Directive,[17] which created a sui generis database right that automatically, just like copyright, arises once a database is established, back the access restrictions up.[18]

There is a counter-movement to such digital enclosure. Open-source efforts such as the Creative Commons License[19] or Wikipedia, the open-source encyclopedia, illustrate this. As Poynder (2003b; p. 33) explains,

> although copyrighted material that's already been assigned to large corporations is probably lost to the commons for a long time, individual creators remain free to produce an infinite number of new works. If in doing so they utilize copyleft, the argument goes, a new commons can be created.

While many million websites utilize such licenses and many software programs are open-source, a large part of valuable content or industrial know-how will not be freely disclosed in such manner. Most industrial companies don't freely disclose their know-how, which often constitutes their competitive advantage. Even if some cutting-edge research is published, the valuable know-how is frequently not disclosed until years later. Although IBM's work on the chemical vapour deposition of copper produced tangible results around 1983, it was not published till 1990.[20]

To further illustrate this we will analyze various examples of free revealing that are mentioned in the user-innovation literature and argue that these cannot be generalized as common commercial behavior. Two examples that von Hippel (2005a) specifically refers to are Allen's (1983) and Nuvolari's (2004) descriptions of collective invention[21] of blast furnaces in Cleveland, England, between 1850 and 1875, and the Cornish steam engine industry around 1800, respectively.

In both cases, success to a significant degree resulted from the free revealing among competitors of enhancements and improvements. It should be acknowledged however that it was a period of time in which IPRs in particular functioned differently, illustrated by the fact that the incremental improvements of the blast furnaces could not have been patented at the time.[22] A further limiting factor to Allen's and Nuvolari's findings is the critical role of complementary assets. In both cases the free revealing made commercial sense as it increased regional profits by attracting mine entrepreneurs to come and invest substantial amounts into the local coke and coal mines. The "specificity of the resulting technical progress to local conditions" (Allen 1983; p. 18) and the "peculiar structure of the Cornish Mining Industry" (Nuvolari 2004; p. 357) do not allow these accounts of free revealing to be generalized. Allen (1983; pp. 20–21) even states that "since 1900 collective invention has probably become less important." Very few complementary assets will be so strong to allow for complete free-revealing. In Henkel's (2006; p. 18) study of free revealing in the embedded Linux industry he found that the hardware of device manufacturers often "does in fact not constitute a sufficiently protective complementary asset." Finally, it has to be clarified that collective invention does not necessarily refer to von Hippel's free revealing and democratization of user-innovation. On the contrary, collective invention can refer to selective innovation as Allen (1983; p. 11) describes via the licensing agreement for the Bessemer plant patent in the US: "The licensing agreement stipulated that the firms practice collective invention among themselves but exclude non-patent holders."

Another often cited example of free revealing is IBM's free revealing of its "copper interconnect" process to competitors that Lim (2004, 2000) reported. Rather than providing support for a general trend towards free revealing this case demonstrates the strategic role that selective revealing can play. IBM developed an innovative process to produce semiconductors that incorporated copper-interconnections instead of aluminium ones. IBM required equipment to implement this process on a production scale and design firms were only willing and able to build such equipment if IBM revealed to them the details of this process. Under these circumstances, IBM freely revealed large amounts of proprietary information to such design firms. In this manner selective revealing enabled IBM's process to became the industry standard, while still allowing IBM to remain market leader.[23] Lim (2000) emphasizes 2 important caveats. Firstly, IBM only revealed to selected equipment manufacturers and generally "continue[d] to guard sensitive process information" as it had done throughout the previous decade of development. Secondly, "it is important to point out that this strategy of sharing technology was only feasible after IBM had established itself as the leader. Otherwise, another firm might have exploited the knowledge to beat it to market."

Morrison et al.'s (2000) finding of free-revealing libraries in Australia should at least to some extent be seen in the light of the public welfare mission of libraries. The fact that "libraries are not direct rivals in the marketplace because they tend to specialize with respect to both geographic coverage and subject matter" (Morrison et al. 2000; p. 1523) also indicates that this case does not refer to typical commercial behavior.

Franke and Shah's (2003) finding of free revealing in sports communities should also be seen with caution. Rather than being motivated commercially, the foremost goal of such hobbyists is to enjoy their sport within their community. Free disclosure of novel designs by users to the community to test is thus normal.[24] Shah (2000; p. 16) mentions the key reasons why such developments were freely disclosed rather than patented:

> Sometimes the technical novelty of the innovation did not rise to the level of being recognized as a patentable innovation . . . sometimes they could not afford the costs of obtaining a patent; sometimes their immediate public use of their innovations made patenting legally impossible;[25] Innovators in this field also did not have the ability to protect their innovations as trade secrets, because innovations are openly displayed during use.

These factors indicate that free-revealing resulted from the circumstances rather than having been the strategy of the relevant firms or individuals. Harhoff et al. (2003; p. 1757) furthermore explain that "the stronger the competition between the user-innovator and other users of the innovation, the larger will be the loss of competitive advantage that the innovator incurs by revealing," and therefore the less likely it is to be revealed. Hienerth (2006) and Baldwin et al. (2006) found proof for the assertion that as the industry matures and commercial motives become increasingly dominant, free-revealing declines substantially in sports communities.

Finally, we need to consider open-source software. Even though OSS has substantially changed the way software is developed, it does not seem to be indicative of a paradigm change in other areas. First of all, the main reason for disclosure frequently is the positive signaling effect such disclosure has on a programmers' ability to obtain a better job.[26] Such effect does not yet seem applicable in other areas such as biotechnology.[27] Secondly, even OSS is often not freely revealed. O'Mahony (2003) explained how software communities frequently do not freely share know-how that may be of strategic importance. Dahlander (2004) found that 4 out of 5 OSS firms in his sample kept the copyright to their entire program, and 3 of the firms closed at least parts of their software. Henkel (2006) documented that firms developing embedded Linux software often delay and restrict diffusion by providing the source code on demand only, or only to customers, after a sizable delay and without any active support. Such practice can result in a considerable advantage in terms of lead time. Furthermore, the majority of the revealed code was generic (85% of responding software companies agreeing to this) and drivers were frequently made available only as loadable binary modules, not as usable source code. "Selective revealing" as Henkel (2006, 2004) refers to it, therefore seems to characterize industrial behavior more appropriately than free revealing as firms are likely to resist completely giving away their most valuable IP—their "crown jewels."[28] As an illustration consider the recently formed eco patent commons. In January of this year inter alia 3 of the world's 25 biggest patent owners—IBM, Sony and Nokia—who together in 2007 were granted 5,311 patents in the US alone,[29] decided to form a commons in which they pledge not to enforce patents on certain environmentally friendly technologies.[30] Given that a total of 31 patents have so far been donated by the corporations in questions—27 by IBM, 1 by Sony, 1 by Nokia and 2 by document management and direct mail company Pitney Bowes[31]—unfortunately the promising initiative currently merely seems to amount to a half-hearted public relations stunt.

The increasing commercialization of the open-source software business further supports this view.[32] According to Graham and Mowery (2004) numerous "firms are engaging in follow-on proprietary innovation using OSS software as a base." Furthermore, alone in the US there are estimated to be around 170,000 active software patents,[33] dwarfing the few patents—around 2,100—that have so far been donated.[34] Huppert (2004) argues that companies patent software specifically because the trend towards open-source software means that they can often no longer appropriate their R&D by relying exclusively on trade-secret law. Feller and Fitzgerald (2002; p. 177) therefore state that "OSS has a large case to be answered before one could be assured of its appropriateness as a model for software development in general."

In this light we would like to mention the private-collective model that von Hippel and von Krogh (2006) use to describe the open-source movement. In their model, innovators benefit from freely revealing as they keep some sources of profit private. The essential distinction between our view and the one of von Hippel and his colleagues is that in our opinion much of the most valuable, the cutting-edge or most up-to-date, know-how will frequently not be freely released and thus remain proprietary. This relates to the point made by Katz and Shapiro (1985) who argue that valuable major innovations will not generally be licensed.

Overall, given that the examples of free revealing refer to over 200 years of inventions, they do not seem to indicate that a general information commons is forming. While in certain niche areas commons may well form, we have some reservations as to whether these will be sufficiently large to enable a large amount of free user-innovation. On the contrary, we believe that strong IPRs and the selective partnering that the permission to innovate entails, can be seen as important parts of Chesbrough's (2003) open-innovation paradigm. Open-innovation refers to the principle that companies should leverage external discoveries and search for companies who have the best business model for a specific technology. Chesbrough's paradigm involves parties actively exchanging technology, often via licenses, to enhance the results for both parties. Rather than referring to a world where IPRs are neglected and become irrelevant, IPRs play a critical role in enabling such "markets for technology." As Arora et al. (2002; p. 117) explain in their homonymous book: "Stronger IPRs can enhance the efficiency of technology transfers, and hence encourage the diffusion of technology."

Many commentators have reemphasized this view. Rivette and Kline (2000; pp. 22, 44, 183) explained that

> intellectual property has become a strategic asset and competitive weapon of enormous value . . . a company will often find that patents may be the only things standing between success and failure. After all, what's to prevent a big company from copying a small start-up's idea, steamrolling over its market space, and eating its entrepreneurial little lunch? Nothing—except, that is for patents!.

Mori (2005; p. 384) confirmed this view for Japan: patents have become "weapons in the struggle to compete." In an IPO Survey on the "Strategic Management in America's Corporations" 92% of respondents indicated that competitive advantage is built on "proprietary technology" (Cockburn and Henderson 2003). Giving away such weapons for free will require clear strategic justifications and will be the exception rather than the rule. Established companies will therefore be very protective of their core technologies and of their later-stage research.[35] Gambardella et al. (2006; p. 9) found that large corporations are generally much more reluctant to license than smaller companies and are "more likely to license fringe or peripheral rather than core technologies." This is because licensing can create new competitors. Arora and Fosfuri (2003) even explained that companies will only license if the "revenue effect," i.e. the gain from licensing, will outweigh the "rent dissipation effect," the loss represented by increased competition. Even if a license is granted, such agreements generally contain terms that require the know-how to remain confidential for years after the agreements' expiry.[36] Granstrand (1999) adds that while in the 1980s Japanese corporations tended to pursue very open licensing policies, granting a license to anyone who could pay for it, since the mid-1990s these companies have become increasingly selective in granting licenses.

In the outsourcing literature the loss of control over one's IP has also been identified as a significant barrier. Howells (1999) for example explains that companies are still reluctant to outsource critical technologies to outside suppliers. Oxley (1999) even provides empirical proof that in countries where IPRs are weak, US firms are more likely to organize an alliance as an equity joint venture rather than a contractual agreement due to the fear of losing control over proprietary know-how.

Why should this be different for free revealing where instead of only one company everybody obtains the legal right or the know-how to be able to freely compete? We therefore assert that the free-revealing advocates underestimate the importance of patents, both as bargaining chips that enable entrance to certain market segments[37] and as tools to increase the time it takes to duplicate an innovation. This can be seen by Mansfield's (1985) study of the leakage of technological know-how that von Hippel and von

Krogh (2006) use to argue that patents are ineffective. In the 100 US companies that he studied, on average "information concerning development decisions is generally in the hands of rivals within about 12 to 18 months . . . and information concerning the detailed nature and operation of a new product or process generally leaks out within about a year" (Mansfield 1985; p. 217). It takes considerably longer however to duplicate patented products and processes than unpatented ones, and this comparison does not refer to know-how which was freely revealed. Levin et al. (1987) demonstrate that it is nearly four times as likely that unpatented products have been duplicated as opposed to patented ones after a three year period.

Using patents therefore clearly prolongs the competitive advantages of a novel product. This is especially significant as the considerable fall in product life cycles over the last decades has in many areas made "lead time" the most important means of protecting the competitive advantages of new or improved processes.[38] Lim (2004; p. 6) summarizes that "in order to recoup the high cost of participating in science, the firm has to be very careful about protecting the domain-specific knowledge it creates, either through defensible patent positions or through secrecy."

Veryzer's (1998b) in-depth case studies of 7 radical innovations developed by Fortune 500 companies indicate that these concerns are equally applicable to the cooperation with users. A main reason for not collaborating with users during the initial stages of the development process was "secrecy concerns and [as] the proprietary nature of these products restricted the amount of information that could be provided to customers about them" (Veryzer 1998b; pp. 142–143).

Finally, we wish to point to another study that von Hippel (1981) refers to. Even before the patenting boom started in the early 1980s, Taylor and Silberston's (1973) *The Economic Impact of the Patent System* did not find support for the prevalence of free revealing. They instead found that over 50% of the companies in their main sample possessed important patents that they would not even license, not to mention freely reveal, except in return for "exceptionally valuable patents or know-how" (Taylor and Silberston 1973; p. 182). They summarized that "there is considerable selectivity of licensing of major patents by British licensors in their home and principal export markets" (Taylor and Silberston 1973; p. 181).

We wanted to illustrate here that while free revealing will be sensible strategic behavior in some situations, its extent does not seem to point to the forming of a substantial commons that could facilitate and foster user-innovation. We will now consider some illustrations of the permission to innovate.

## SCIENCE AND UNIVERSITIES

The key area that is used to feed the information commons, namely academic science, has been changing dramatically over the last decades. In the late 1980s Eisenberg (1989; p. 1086) already pointed to the gradual establishment of a permission culture in this realm: "Patent protection increasingly threatens to encroach on the domain of research science." The effect of this patent expansion on university researchers has been especially pronounced as the whole research environment has become increasingly proprietary.

While researchers never used to be concerned about intellectual property issues and could simply focus on their research endeavors,[39] over the last 2 decades they have had to rethink and restructure their organizational processes. Before starting their work researchers increasingly have to check the patent landscape to prevent their results being blocked by the patent thicket. While theoretically, research exemptions are supposed to protect researchers from patent infringement, in practice in many countries these have provided insufficient protection. This has been especially pronounced in the US, where cases such as *Madey*[40] and *Embrex*[41] have substantially narrowed the research exemption. In *Madey,* Duke University was unable to rely on the research exemption as the patents in question were in furtherance of the university's normal business. The US National Research Council (2004; p. 108) explained that *Madey* could reasonably be interpreted to hold "that formal research enjoys no absolute protection from infringement liability regardless of the institutional venue, the purpose of the inquiry, the origin of the patented inventions, or the use that is made of them." Saunders (2003; p. 268) concurred and indicated that we may be moving to a permission culture where university researchers can no longer "stand firmly on the shoulders of giants" but rather have to "rent space on those shoulders."

To navigate the IPR thicket and negotiate potentially necessary licenses, researchers, before being able commence their research, will increasingly require the assistance of patent attorneys or of technology transfer managers to undertake 'freedom to operate' patent reviews. Another difficulty that both university researchers and other users, especially in the medical realm, often face is the lack of funds to purchase leading-edge equipment. MTAs and reach-through license agreements (RTLAs) can offer a solution. Such agreements can allow researchers with limited funds to obtain access to expensive patented machinery or research tools with which they can experiment without initially having to worry about high financing costs.[42] Payment is generally conditional on research results. A typical MTA would for instance provide the manufacturer with an automatic non-royalty-bearing license to the research findings made with the machine in question.[43] RTLAs, which are especially common in biotechnology, require licensees to "pay royalties derived from the revenues of any product developed using the [research] tool" (Jaffe and Lerner 2004; p. 63). As an illustration of their prevalence, Razgaitis (2004), in a study of US and Canadian license transactions, found that 31% of out-license agreements and 36% of in-license agreements contained reach-through provisions. Intel for instance underwrites a substantial amount of university research in order to obtain royalty-free access to patented technologies that result from the funded research.[44]

While in theory such licenses can be an ideal way of meeting the interests of both users and manufacturers, it has to be acknowledged that MTAs can "tightly restrict [how] researchers use [the licensed] materials" and machines (Rai 1999; p. 12). William Lacy (2000; p. 77) of the University of California at Davis even stated that MTAs may "impede or limit the pace and direction of scientific efforts, restrict scientific communication or undermine an academic scientist's ability to carry out research." Heller and Eisenberg (1998; p. 699) therefore restated their previously voiced fear that the use of RTLAs and MTAs can lead to the emergence of the "anti-commons," in which due to multiple owners each having the right to exclude "no one has an effective privilege of use." Walsh et al. (2003) found that in practice in the field of biomedical innovation an average of 6–12 patents had to be seriously addressed and 3–6 patents licenses had to be negotiated. Jaffe and Lerner (2004; p. 63) finally stated that "the need to secure multiple reach-through licenses can impose a large tax on the innovation process, and thereby significantly retard innovation."

These contentions are supported by some empirical evidence. A study conducted in 1997 and 1998 by the National Institute of Health's Working Group on Research Tools concluded that "many scientists and institutions involved in biomedical research are frustrated by growing difficulties and delays in negotiating the terms of access to research tools" (NIH 1998; p. 1). Scientists wait months or even years to carry out experiments while their institutions try to negotiate the terms of MTAs, database access agreements, and patent license agreements. Even pharmaceutical companies have complained that access to basic research tools has often become burdensome.[45] A survey of public plant breeders at 25 US universities in regard to 41 crops similarly found that many scientists had experienced difficulties obtaining proprietary genetic stocks from private companies and that this had adversely interfered with their research.[46]

This trend towards the restraint of the freedom to research is accompanied by the increasing commercialization of universities. Especially the research departments of US universities increasingly look like private research institutions.[47] Since the passing of the Bayh-Dole Act in 1980 allowed the title of an invention to vest in the university itself, patenting has boomed in this area.[48] Their links to and dependence on industry have become so considerable that Derek Bok (2003), former President of Harvard University, stated that in many areas, such as medicine, this process is already irreversible. In the early 1990s corporations for example already supplied over 45% of the funding for academic biotechnology.[49] One of the most publicized examples is the contract between Novartis and the Department of Plant and Microbial Biology at the University of California, Berkeley. Novartis paid $25 million over 5 years (around 30–40% of the department's entire budget) for the right to review in advance all proposed publications that are based on their funding, the right to demand the university to file a patent on any findings, as well as the first right to negotiate a license to such patents. Additionally, 2 out of 5 seats on the committee that decides how to allocate research funds were given to Novartis.[50]

In their book *Academic Capitalism*, Slaughter and Leslie (1997) argue that there has been a clear shift in North American, Australian and British universities away from basic research to targeted or commercial research. The AUTM Licensing Survey (2007) further supports this proposition: Research expenditure, patent applications and license agreements have all risen sharply. Holding equity positions in spin-offs has become increasingly common for researchers, thereby making their behavior increasingly entrepreneurial and proprietary. Jennifer Washburn (2005) in her *University, Inc., The Corporate Corruption of Higher Education* further explains how US universities of today have assumed "a new proprietary culture more akin to the business world." Indeed even MTAs from one university to another have included restrictive clauses including ones which require the assignment of intellectual property rights to those developments that arise from the use of such tool.[51] In one particularly worrisome case researchers at the University of Utah, who had discovered an important gene responsible for hereditary breast cancer, obtained patents and formed Myriad Genetics for commercialization purposes. Originally Myriad not only restricted other scientists from using the gene but also threatened legal action against these. Although Myriad subsequently extended licenses to the leading medical schools, universities and hospitals, giving these the right to use the gene for their breast cancer research, such licenses were so restrictive—they for instance prohibited clinical testing—that various scientists abandoned their research programs.[53] The AUTM Survey's (2007) found that 39% of university licensing agreements in 2003 were exclusive and that only 14% of licenses went to start-up companies, which further indicates that free university revealing has at least in some areas become endangered.

We have seen, as the Golden Rice case illustrates, that to some extent a proprietary permission culture that endangers the free exchange of ideas, already exists in the realm of science. If this is the case in the traditionally open world of science it is likely to be even worse in the traditionally secretive realm of technology.[54] Simply comparing the size of US industrial research, $194 billion of R&D in 2002, around 70% of the national total, to the $33 billion spent by US universities, i.e. 12% of the national total,[55] further indicates that a free and open information commons that could lead to the democratization of user-innovation is currently not forming.

## THE PERMISSION TO INNOVATE IN OTHER REALMS

Unfortunately, such a permission culture extends far beyond the realm of university researchers. Two other related areas that can be characterized by the permission to innovate are the pharmaceutical and the biotechnological industries. Due to the immense cost of R&D[56] and the wide variety of skills that are needed to succeed in both of these realms a web of licenses has formed that enables different companies to innovate by cooperating with one another.[57] Without obtaining such permissions to innovate innovation in these realms will often be impossible.

The realm of medical instruments can also be seen from a permission to innovate perspective. Joint user-manufacturer innovation has become the industrial norm in this area.[58] Through tools such as MTAs manufacturers selectively collaborate with leading-edge users:

> The manufacturer looks for users, capable of handling unfinished equipment and with ideas for improvements and new applications. Such criteria rules out most of the potential users of the result of this process, i.e. the improved product. This only leaves a small group of experimenting technology leaders as candidates. (Lotz 1992; p. 89)

Once made, such collaborations enable manufacturers to benefit from the sticky know-how that doctors possess, namely the understanding of their patient's health problems and how to relieve these. Doctors on the other hand can obtain access to leading-edge technology and have the possibility of creating a reputation for themselves as technology leaders.[59] Manufacturers will also generally be needed at the manufacturing and commercialization stages and to cope with the IPR and regulatory thickets.

In other areas commercial strategy has led to selective co-creating of value with customers.[60] IBM for example, in its FOAK (First of a Kind) program,[61] contracts with a customer to solve a commercially important problem. IBM dedicates its own research staff to work for an extended period at the customer's site. While the customer obtains a solution to his problem, IBM can "expose its own internal research staff to cutting-edge problems at its customers' site" and thereby secure any resulting IPRs (Chesbrough 2003; p. 110). The testing of Microsoft's operating systems such as Windows 2000, XP or recently Vista can also be seen from this perspective. Here thousands of programmers were given the opportunity to co-develop such software subject to clear limitations. Not only was the source-code not revealed, but at least in case of Windows Vista, the beta-versions were time-limited to expire at a predetermined date. Nevertheless both sides benefited from such controlled co-development: Prahalad and Ramaswamy (2000) estimated that consumers invested some $ 500 million worth of time, effort and fees into the co-development, largely via feedback, of Windows 2000.

Extreme Programming (XP) is another example of selectively co-creating value with customers. XP is an agile method of software development that involves the "incremental, iterative development of sequenced small improvements" through a feedback loop in which the developments are continuously passed on to selected customers (Gassmann et al. 2006; p. 53). On the basis of their feedback the product is modified and then passed on to the customer again. Through such close collaboration, products can allegedly be developed more effectively due to the ability of "integrating technical and market knowledge derived directly from the customer" (Gassmann et al. 2006; p. 62). This clearly is a promising example of the "permission to innovate" concept as only a few select customers will be granted access to the highly valuable corporate know-how. Often these will be clients for whom the technology is being tailor-made. The Swiss elevator company Schindler for example often closely collaborates with its customers, in one case to create a sophisticated access control system for their elevators in a clients' office building. Another example is Mitsubishi Electric Corporation which calls this co-innovating process "concurrent engineering" (Branscomb and Kodama 1993; pp. 32–33). In one case of applying computer systems to energy management systems, the complexity of the approximately 2 million lines of code was dealt with through a close and open collaboration with the client.

A final software example is Palm OS, the operating system for handheld devices. While Palm has given away its source code to over 150,000 external programmers, who have created over 10,000 third-party applications, they not only have to be registered with Palm to develop applications but more importantly no third party is allowed to touch the operating system itself. The firm's policy is that anyone who modifies or tinkers with it in any way will be sued.[62] When Palm OS is licensed to competitors such as Sony and Handspring, these only receive licenses to use the software not to modify it. The control over its product is crucial to sustain Palm's competitive advantage, just as it will also be of considerable importance for many other companies.

## PUTTING THE PERMISSION TO INNOVATE INTO PERSPECTIVE

The paradigm of permission to innovate entails selective user-innovation and should be seen as part of the broader open-innovation paradigm. We have seen that as companies are interested in sustaining their competitive positions and their monopoly rents, open and free user-innovation will not always be in their interest. Companies may therefore attempt to prohibit such activities. Whether or not this will be advantageous to the company in question will depend on the intricacies of the situation. In certain situations embracing innovative users may certainly be advantageous. While in either case users will continue to play some role in the innovation process, their ability to innovate may increasingly depend on external help and permission. Access to leading-edge know-how or machinery may not feasibly be attainable in another manner. Our examples indicate that rather than completely 'free' user-innovation being or becoming the industrial norm, selective user-innovation seems to be a more accurate characterization for numerous industries.

We clearly need to be careful about these findings. Further empirical research is necessary to determine in which areas our or von Hippel's characterization is more accurate. Based on existing data it is difficult to precisely compare user-innovation throughout different time periods and make accurate predictions for the future. Such analysis would be the role of economic historians and due to the tremendous amount of factors that would have to be considered a very challenging task indeed. Clearly, there is some truth in both views: Our world will be characterized by both open and closed innovation, free revealing and know-how-hogging, user-, manufacturer- and third-party-innovation.

# D.2 Licenses as Sources of Innovation

In the previous sections the argument was made that both the right and the ability to innovate were often necessary components of the innovation process that many users did not possess. In this section we would like to fill an acute hole in the sources of innovation literature and demonstrate that licenses can be an important source of innovation. We will use this realization to strike the link with user-innovation theory. The argument will be made that manufacturers can use licenses to collaborate with certain (advanced) users. This can be a sensible commercial course of action as Lüthje and Herstatt (2004; p. 556) explain: "The involvement of innovating customers in the early-stages of the innovation process is a promising strategy for manufacturers with a current need for innovations."

## THE IMPORTANCE OF LICENSING

Licensing is the act of granting another party permission to use your intellectual property.[63] More precisely, it refers to a contractual arrangement where one party, the *licensor*, authorizes another, the *licensee*, to use an object, his know-how or IPR for a certain agreed purpose, generally for the production and commercialization of a product or service.[64] In this simplified view the licensee is the recipient of a right to use a technology (patent plus know-how) granted by the licensor. The difference to a sale is that the licensor remains in control of some rights in respect to the technology.

Licensing is an inherently versatile strategic tool that is not only fundamental for marketing purposes,[65] but that can also be used to enhance both sales and profits, to enter foreign markets, to outsource R&D, to get access to the resources of another company or to pursue joint research efforts with another company. In the last half of the 20th century licensing has become "a vibrant and extremely important part of business and society" (Razgaitis 2004; p. 140). Bills (2004; p.94) explains that "licence agreements, if they are negotiated to meet the strategic objectives of both the licensor and licensee, can provide a win-win outcome for both parties." While there are a wide variety of different types of licensing arrangements ranging from brand licensing to fashion licensing, we will focus on technology licensing.

Even among technology licenses there are substantial differences. The main distinctions between different types of licenses for our purpose are based on exclusivity, payment structure, and the exchange of know-how and technology. Exclusivity refers to the extent as to which the licensee is the sole recipient of the right in question or whether there are numerous other licensees. Broadly speaking there are three kinds of licenses in this category. Exclusive licenses, where the licensee is the only party who possess the right to the subject matter of the contract, semi-exclusive licenses, where the licensor also possess the right to use the technology in question, and non-exclusive licenses, where there are numerous parties who possess rights to the technology. It does have to be pointed out that this refers merely to the patent as a national right. Consequently, an exclusive license can be nationally, regionally or globally exclusive. Apart from such territorial exclusivity, the field of use specification will also be crucial.[66] Payment structures can be monetary or non-monetary such as technology and know-how exchanges through cross-licenses and grant-backs. Cross-licenses are ex-ante agreements where it is agreed to exchange rights to different but usually related technologies. Grant-backs are obligations on the licensee to return in some way to the licensor know-how or IPRs generated in the use of the licensed technology.[67] This can involve fully assigning such rights or merely granting the licensor a non-exclusive license. The reverse of this is the follow-up clause, where the licensor has this obligation. In mutual exchange clauses both parties must license future improvements to the other:[68]

Licensing is one of the most important instruments for technology transfer between firms and one of the most common inter-firm contractual agreements:[69]

> From the beginning of the [US] Republic, the established business practices included the licensing of intellectual property. Patents have been licensed since the Constitution was written; trade secrets have been licensed since the beginning of the Industrial Revolution; and software has been licensed since it was invented.[70] (Ring 2005; p. 25)

Today, licensing is a widely-spread strategy in the metals, pharmaceuticals, chemicals, foodstuffs, tobacco, entertainment, leisure, software, electronics, computers and biotechnological industries.[71] It is of especially high importance in the realms of high-technology.[72] Anand and Khanna (2000; p. 103) therefore emphasized that "the importance of understanding the economics of licensing seems certain to increase." A benchmark survey undertaken by the US Department of Commerce suggests that during the 1990s foreign licensing by US companies grew at rates exceeding 11% per annum and therefore faster than both trade and foreign-direct investment.[73] Duysters et al. (2000) explain that the share of non-equity alliances, such as licenses, have increased substantially from below 10% in 1970 to nearly 90% in 1997. In an OECD questionnaire 60% of respondents also reported increased inward and outward technology licensing over the last decade.[74] The estimates of the worldwide license market or "market for technology" vary from $35 billion to $150 billion[75] per annum.

What are the reasons for this increase in technology licensing? From a macroeconomic perspective, the growth in licensing has to be seen in line with the overall growth of alliances.[76] Increased international competition due to the homogenization of international markets combined with the increased differentiation of products, the trends towards specialization[77] and concentration on core competences, the increased strengthening of IPRs[78] and their gradual international harmonization and more effective enforcement have played a substantial role in this increase. The rising strategic importance of speed to market, the acceleration of technical change, the growing diversity and complexity of knowledge sources, an increasingly mobile and transient workforce, the emergence and spread of ICTs, and escalating R&D costs are further important reasons for such rise.[79]

A recent study by the LES (USA and Canada) nevertheless found that "only a small portion of what is believed to be licensable IP actually is licensed" (Razgaitis 2004; p. 139). For instance, the survey suggests that only about one of eight opportunities believed to be licensable became part of serious negotiations. Gambardella et al. (2006; p. 3) therefore conclude that "there is room for increasing the actual rate of technology licensing." Arora et al. (2002; p. 8) in their influential work *Markets for Technology* also pointed out that currently "the licensing market is less developed than socially desirable." Nearly 70% of respondents to a recent OECD study expected both their inward and outward licensing to increase in the future to gain access to complementary technology, reduce R&D costs and innovation time.[80]

Currently, licensing with respect to users plays a small role. Users who have developed a prototype for example often grant licenses to manufacturers to complete the development process and market the technology[81] as was the case with CDENT. Granting licenses to users however to allow these to overcome the barriers that we discussed above is currently very rare.

## LICENSING: A TOOL TO GENERATE INNOVATIONS?

Before applying licenses to users in more depth, we will see how licenses can be used to generate innovations. To summarize this literature, the prevailing view is that licenses are primarily technology marketing instruments.[82] The majority of these commentators implicitly do not seem to regard licenses as a potential source of subsequent innovations. A close scrutiny of innovations that result from licensees is lacking completely. Many leading works on innovation such as the *Oxford Dictionary of Innovation,* Trott's textbook *Innovation Management and New product development* or von Hippel's *The Sources of Innovation* fail to mention licensees as a potential source of innovations.

Numerous studies have however shown that licensing allows licensees to fill their skill-gap with effectively sourcing-in external technology at a low cost and risk.[83] Licensing was found to be a complement to internal R&D.[84] Martinez et al. (2004; p. 97) further found that across a wide variety of industries the two most important factors influencing licensing strategies were the acceleration of the innovation process and the access of complementary technology. We should therefore turn our perspective by 180° and ask whether licensing can be seen as effective R&D outsourcing vehicles.

We are not the first to have thought in this manner. In a world in which R&D, due to the rising technological complexity, has become very expensive and often insufficiently cost-effective,[85] it becomes necessary to find new ways of optimising R&D. Ardisson and Bidault (1986; p. 60) therefore suggest that licenses can be a suitable tool to form international networks. While their particular focus was mainly on market entry, we can see from the above discussion that this could also refer to mutual innovation efforts.[86] Brinton (1993; p. 438) in this respect suggests a "hub and spoke" model under which the licensor develops the core technology (the hub) while granting permission (through licenses with grant-back clauses) to allow other companies to develop the specific applications (the spokes). This can enable a party to market and improve their core technology via external partners without losing control over it. It allows licensors to benefit from the different idiosyncratic innovative capabilities of different licensees and "tap into a stream of new product and process ideas and quality improvement potentials it could not possibly generate itself" (Quinn and Hilmer 1994; pp. 52–53).

Numerous studies have implicitly considered licensing agreements to be alternatives to research joint ventures.[87] In this way the potential of licensees to not only commercialize a technology but also to develop it has indirectly been recognized. Wilson (1977; p. 172) has argued that "Licensing of inventions is a potentially important tool for achieving joint maximization of R&D." Matsunaga (1977) further explained that non-modification clauses were detrimental to both parties as they stifled the licensees' innovative potential. Comerford (1984; p. 46) added that "many companies actually improve on the products they have licensed."

Bessy and Brousseau (1998), in a survey based on 46 of the most representative licensing agreements taken from a sample of the leading French firms, found that 65.2% of all these contracts contained grant-back clauses.[88] In 8.3% of these cases the licensor "relied on the licensee to provide him/her with additional developments" (Bessy and Brousseau 2000; p. 71). In Bessy et al. (2002) grant-backs were included in 46.9% of the licenses in the sample. 20 years earlier, when analyzing a sample of US, UK and Canadian firms, Caves et al. (1983; p. 260) obtained a figure of 43%. A previous study of 428 LES members found that nearly 80% of licensing agreements contained a provision for improvements.[89] Vickery (1988) in a study of 119 LES members found that 17.5% of companies even held grant-backs to be the most important element in a licensing agreement. Despite of these realizations, and the inclusion of grant-back clauses in many licensing contracts,[90] no study has further demonstrated the innovative potential of licensees. In this respect we believe that is important to move away from a static analysis of looking at the mere inclusion of contractual clauses to a dynamic analysis of the actual innovations generated. We will therefore briefly provide three empirical examples of innovations generated by licensees. In a forthcoming set of articles to appear in the LES Nouvelles, "Licensing to Innovate," these and many others (e.g. Japan, the steam engine, the Jet ski, insulin or the sewing machine) are discussed in more depth. It is crucial to note that when we refer to improvements, we do not just mean the adaptation or modification of the licensed product to local conditions. This is a task which licensees are generally believed to do very well.[91] Improvements for our purposes rather refer to widely applicable changes that can also be useful for the licensor and which thereby enhance the value of the licensed technology.

We previously mentioned that CDENT is gradually revolutionizing dental care. We will now briefly discuss its development and show how a license transaction was the basis of three subsequent generations of innovations that revolutionized the dental industry.

Back in the early 1980s, the dental industry was still very fragmented and consisted of many conventional low-tech enterprises. High-technology had not yet made its way into the field. At the time Professor Werner Mörmann of the Dental School of the University of Zurich and Dr. Marco Brandestini, an engineer and former head of development of a dental equipment company, developed the idea of creating a computer-assisted design and manufacturing tool for the dental industry. After some tinkering they were able to patent their idea. To commercialize these they formed "Brains" (i.e. Brandestini Instruments). Within the following years a workable prototype was constructed and in 1985 the first patient was test-treated at the University of Zurich. Brains, with the intention of producing the device, purchased kits to construct 25 machines and commenced to build these by hand. The inventors quickly realized that they lacked the production and marketing know-how as well as the finances to adequately commercialize the technology themselves.

Fortunately for all parties involved they were able to sign a globally exclusive licensing contract with a prominent medical device company. The licensee marginally adapted the product to make it ready for serial production and within a year commercialized CDENT 1 in 1987.

The goal of CDENT 1 was to produce inlays, i.e. solid fillings such as cavities. Overall due to the limitation of the machines its market success was very limited, with around 250 machines having been sold annually between 1987–1994. CDENT 1 was a unique concept as it enabled the dentist to produce inlays himself in one relatively short session for which he would otherwise need a laboratory that may take various days.[92] The technical realization of the machine however according to our interviews under accomplished user requirements. The early launch of the product allowed the licensee to gather a substantial amount of market know-how. After years of further R&D a substantially improved version of CDENT was introduced in 1994.[93] This device could also handle front teeth[94] and 800 machines were soon sold annually.

During the next 6 years further sums where invested into product improvement. While in 1999 the licensee finally purchased the patents, the vast majority of the main improvements for the third generation of CDENT where made as a licensee. CDENT 3, a sophisticated Windows NT-based CAD/CAM system, was introduced in 2000. While originally CDENT could only produce single- and dual-surface inlays, CDENT 3 could create multi-surface inlays, single onlays, veneers, and posterior and anterior crowns.[95]

To summarize, the original license from Brains to Siemens resulted in a "whole family of products" or even in the birth "of a whole new industry" as the current CEO of the manufacturer describes it. Today over a third of the original licensee's sales come from CDENT products (Annual Report 2004/2005; p. 22).

## Photocopier

One of the most significant innovations for the modern office, the photocopier, was developed due to the innovative efforts of a licensee. In 1942, the American Chester Carlson was granted patent #2,297,691 for "electron photography." As his development was still at a very early stage—Carlson was merely able to copy objects written on a glass plate onto waxed paper—the inventor desired a corporate partner. After a long struggle he finally managed to negotiate a license with Battelle Memorial Institute that agreed to develop the idea and to grant Carlson royalties.[96]

In 1947, Haloid, at that time a small New York based organization manufacturing and selling photographic paper approached Battelle to obtain a license to develop and market a copying machine based on Carson's technology, which Battelle had already improved upon. Haloid started to spend a quarter of their annual net income for over ten years on developing a workable office product from the underlying technology, essentially betting the company's future on it.[97] The first small success of Haloid's effort was the development of the Haloid 1385 in 1951, the first electro-static copier marketed in the US, which was manually operated and took several minutes to make each copy.[98]

Only in 1959, 21 years after Carlson first filed the patent for the underlying technology, was the first plain-paper office copier announced and brought to market.[99] This automatic machine that produced seven copies a minute was the key to Xerox's, the new name of Haloid, ultimate success.[100] In the same year Xerox finally bought all of the worldwide patent rights on xenography from Battelle. By the early 1970s, the improvements of these patents and Xerox's marketing had made the company so successful that Xerox was sued by the Federal Trade Commission for monopolizing the market for photocopiers[101]

## Wind Energy

The development of the global wind energy industry has also been substantially influenced by licensing. Both the world market leader Vestas and its biggest competitor, Gamesa Eolica developed their original technology mainly from a license. In the mid 1970s a Danish mechanic, Karl-Erik Jorgensen, improved upon the previously existing wind turbine designs and built prototypes using glass-fibre blades and an electric motor that turned the turbine to face the wind. He subsequently sold a license to use his improvements to Vestas. Through utilizing and improving upon Jorgenson's design Vestas has since become the world's largest wind turbine manufacturer.[102]

The Spanish company Gamesa Eolica, the global number 2[103], ironically developed its technology based on an exclusive license from Vestas.[104] Gamesa quickly improved upon the technology it had licensed-in and today is no longer "a net technology receiver" but instead transfers its own proprietary technology to others (Gamesa 2004; p. 35). The importance of licenses in the wind energy industry in respect of the development of both the underlying technologies and the organizations in question remains high, with both Gamesa and Vestas having further licensed their technologies to Asian companies.[105]

# D.3 Licensing to Innovate

## An Alternate Instrument for User-Manufacturer Co-operation

Prahalad and Ramaswamy (2004; pp. 4, 12) argue that companies can no longer autonomously design and develop products as the future of corporate competition will involve "an individual-centred [not corporate centred] co-creation of value between consumers and companies." Through such close collaborations with users manufacturers can stay at the leading edge of an industry. These can enable manufacturers to potentially obtain high quality contributions from users and to obtain a better understanding of user needs and how these can be satisfied. Translating user needs, in functional terms, into products is often a very difficult process, especially as customers frequently cannot express their exact needs unless they can assess a product in real-life situations.[106] As users possess specialized knowledge about the environment in which the products will be used, such co-developments offer the "opportunity for the mutual adaptation of both technology and user work environment" (Leonard 1998; p. 104). Voss' (1985) study of the software industry confirmed that intensively studying users' needs substantially increased the likelihood of developing successful software products.

Companies may also want to collaborate with advanced users for sociological reasons. Stigmatizing tinkering users as criminals is "a recipe for a continuous war" between such users and the company in question (Mollick 2004; p. 94). It may therefore be wise to integrate the best deviant users into product development and turn their destructive ideas into progressive ones.[107] The results of the CDENT study confirm this view. Some companies such as Lego, Valve or id Software have successfully turned a significant section of the previously rebellious user-innovation community attached to their industry into co-developers of product extensions.[108] Users have for instance developed a new improved operating system for Lego's Mindstorm robots.[109]

There are various ways in which manufacturers can deal with the innovative potential of users. Von Hippel (2005b; p. 74) outlined three options. Firstly, manufacturers can actively seek innovations developed by lead-users which they can then integrate into their product development activities (i.e. the lead-user method). The second option goes considerably further in that manufacturers can "draw users into joint design interactions by providing them with 'toolkits for user-innovation'" von Hippel (2005b; p. 74). The third option, allowing users to become manufacturers themselves, is not likely to be one favored by most manufactures.

We will now briefly explore and expand upon the second option. The concept of granting toolkits for users has become increasingly popular in recent years. The theory states that toolkits can transfer "need-related product development tasks from manufacturers to users, and equip the users with tools to carry out those tasks" (von Hippel and Katz 2002; p. 822). The term "toolkits for user-innovation and design" that many commentators employ is however at a minimum semantically uncertain. In their articles Schreier (2005) or Franke and Piller (2004) discuss the concept of mass-customization, tailoring the design of watches, mobile phone covers, t-shirts or shoes to the needs of the user by allowing these to participate in the products' design. The changing of the color, the choice of alternate predefined materials or a user-designed logo amount to design contributions but not to product innovation. While the boundary between innovation and design may be fuzzy, simply matching and mixing design opportunities from a large sample of choices is merely design customization.[110]

While both Franke and Piller (2004) and Schreier (2005) found that such personally designed products lead to an increased willingness to pay of over 100%, it has to be noted that most products are more difficult to design than t-shirts, shoes or scarves. Most people would not for instance possess the skills to design their own vacuum cleaner or tennis racket. In this respect we should distinguish between functional and visual design. The later refers to the external appearance which may be tailor-made by users in many situations. Functional design, which involves amending the functionalities of a product, is much more complicated. The current toolkits rather than referring to user-innovation largely refer to visual design. Schreier (2005; p. 24) confirms this view: "Existing toolkits are limited by the constraint that the result is merely a range of personalized products (and no real innovation)" (own translation).

This imprecise semantic combination of innovation and design toolkits is misleading and needs to be corrected. While the line between mass-customization and innovation toolkits will be a fluid one, it should be recognized that there are various types of toolkits. The literature has already pointed out the difference between high-end and low-end toolkits. High end refers to sophisticated toolkits with a focus on the development of innovative new products. Low end toolkits on the other hand offer simple personalization/customization opportunities for standard products. A further critical distinction is between basic and expert toolkits. While the former can be used by everybody, the later require a higher degree of sophistication and therefore only appeal to advanced user groups.

The question of "how would the rights owners know who should get permission to circumvent their technology?" that Varian (2002) posed, while very important, is beyond the scope of this work. It is clear that finding an appropriate partner and successfully negotiating a suitable agreement can be a long, cumbersome and expensive process.[111] Nevertheless, we will briefly mention a few important considerations. Clearly it is far easier for manufacturing firms to identify users who desire to use a technology rather than those few who are interested and capable of active development contribution. Leonard (1998) and Franke et al. (2006) mention some of the key criteria for identifying users for innovative purposes:

- possess a high level of skills and know-how
- are willing to contribute
- have a high benefit from finding a solution
- possess community-based resources

Rogers (2003; pp. 282–283), summing up the diffusion of innovation literature, stated that "innovators," i.e. the very first adopters of a new technology, possessed "substantial financial resources," "the ability to understand and apply complex technical knowledge," and a risk-taking attitude, as wellas being interconnected within the respective social system. Westphal et al. (1985) and Mordhorst (1994) both emphasized the importance of previously existing know-how and adequately trained staff in order to adequately assimilate and subsequently improve a technology. Clearly the exact criteria will depend on the industry concerned and the problem at hand. Manufacturers should attempt to collaborate with users that can provide them with complementary skills and resources. The literature has so far identified four principal methods that can be used to select lead-users:

1. *Screening:* By sending out a questionnaire to a large group of users, who are identified largely through freely available information from customer data banks, customer complaints or external customer reports, and inviting those that fulfil certain criteria to an idea-and concept generation workshop.[112] This approach is practical when the number of customers in the market is overseeable and a complete screening of all users is possible.
2. *Pyramiding:* As experienced users tend to know even more sophisticated users, a relatively small amount of experienced users are asked whom they believe has the most expertise in a field. By climbing up this ladder of experience, eventually global leading-edge users can be identified.[113] A significant advantage of this method is that users in analogous fields in which similar challenges are present can be identified. As an example consider a lead-user project in the field of medical imaging to recognise small tumors. Apart from leading radiologists experts from the military field where also identified as lead-users. In this manner the common processs of pattern recognition software to analyse satellite images was transferred to medical imaging, where the focus had previously always been on improving resolution.

3. *Signalling/self-selection:* This relies on self-selection and virtual communities. By signalling to a virtual community that one is looking for leading-edge users with whom to develop new products, one can assume that the most knowledgeable and skilled users self-select themselves as these have the most interest to participate in such a project.[114]
4. *Internet-based idea-competition:* Customers are invited to submit innovative ideas to the manufacturer and as an incentive prizes are awarded to the best submissions.[115]

A software program has been specifically designed to identify lead-users. While the results so far have been unsatisfactory, future improvements, especially if the utilising company more actively triangulates users' self-assessments, may lead to more successful results. Lüthje and Herstatt (2004; p. 560) report that manufacturers can "use the information on user's characteristics to predict the specific application area and innovation solution type the users will probably develop."

We should expect there to be a very high correlation between high-end and expert toolkits, as experts are required to appropriately use advanced toolkits. Prügl and Schreier (2006; p. 247) describe that "whereas low-end toolkits which focus on customization (by allowing minor product adaptations) could serve as a gateway, sophisticated high-end toolkits with more innovative design freedom might be offered to more experienced users." This idea that different toolkits could be employed for different types of users can be expanded to consider a further crucial distinction: The distinction between granting toolkits to *all* users (or all who want to pay for it) or simply to a *select* group of "lead-users."

The reason why such a distinction is necessary is that not all companies will be keen to grant what Prügl and Schreier (2006; p. 246) call an "open invitation to innovate," with the accompanying potentially detrimental consequences of facilitating innovation by competitors. The Vestas/Gamesa case illustrates that the granting of a license can create a strong competitor.[116] Choi (2002) explains how Ampex was the first company to introduce a commercially viable videotape recorder and how through wide licensing its technology without grant-back clauses it quickly lost its technological edge. Sony and JVC instead succeeded in mass-marketing the VCR. Another illustrative story is that of Swiss diesel engine designer Sulzer, who was taken over by its licensee Wärtsila of Finland.[117]

Hope (2004; pp. 213–214) therefore indicates that this may not necessarily be in the interests of manufacturers: "As customers and independent tool developers, eventually learn to design toolkits that are not tied to the production facilities of a particular manufacturer." In this light Douthwaite's (2002) suggests that users should self-select themselves by demonstrating that they are willing to commit a considerable amount of resources, especially in terms of time, into the innovation process and possess some advanced know-how. There is little point in collaborating with mediocre users if new innovations are intended to be derived from these rather than being created for them. In the late 1970s, Telesio (1977) explained that licensees will frequently struggle to obtain access to cutting-edge technologies in numerous fields if they cannot offer any technology in return or do not possess a strong R&D department. Why should users therefore receive leading-edge technology for free without any commitments in return?

We therefore believe that intellectual property issues will play an important role in many toolkits, especially in the more advanced ones. The inclusion of grant-back clauses can safeguard the licensor by discouraging the licensee to invent around his technology.[118] Being able to exert some control over the developments that users may make with the toolkits, such as by obliging these to assign the rights to their developments to the manufacturer of the toolkit, will often be an important factor. To summarize the more proprietary know-how is revealed or access to expensive equipment is granted (i) the higher the likelihood that strict contractual obligations will be imposed on the users and (ii) the smaller the group of recipients will be.

Those toolkits that really deserve the name toolkits for user-innovation, which we will refer to as licenses to innovate, will include the critical tools, know-how and any other resources (finances or staff) that users need to effectively tinker and innovate, in the frequent situation when they are unavailable in the commons. In this manner the license can enable users to overcome their "lack of technological competence" in certain areas (Lotz 1992; p. 52). Levin et al. (1987) found that most parties need some help to reach the frontier from which innovation is possible and that apart from independent R&D, licensing is by far the most important method of learning about new products and processes. The majority of licensing agreements contain significant amount of know-how transfers to the licensee—Vickery (1988) in a survey of international licensing transactions found 67%. The possession of both such technological know-how, combined with the acumen generated in using the technology should help users obtain the ability to innovate.

The license to innovate will also grant users an assurance that they will not be sued for tinkering with the manufacturer's product, which of course is always subject to the rights of third parties. This right to tinker is a substantial advantage over the ordinary user, who generally merely holds warranty/guarantee rights. Unlike the average license to use a product the license to innovate does not contain a non-modification clause which contractually prohibits licensees tinkering.[119] To illustrate take the average (non-open source) software license that entitles the purchaser to use the software according to its terms, but generally prohibits any form of modification. Further consider the relationship between Sun Microsystems and Microsoft. The latter had licensed the right to use Java in the mid 1990s and modified it so that programmers could write applications that ran better on windows. As these modifications however were in breach of the relevant licensing agreement Microsoft had to pay Sun $20 million to settle the dispute.[120] We therefore define that the license to innovate contains both the legal permission to innovate and the transfer of know-how and other resources to enable a licensee to improve the licensed technology. In this respect it is very similar to grant-back licensing. In return for granting the user special know-how and tools, the user agrees to transfer at least a part if not all of the IPRs that he may generate to the licensor.

| License to Innovate = |
|---|
| RIGHT (Legal Permission) + ABILITY (Technological + Economic) |

The main reason for our assertion that licensees are a likely source of innovation is that they are in an ideal position to combine market-pull ideas, those which are generated by the marketplace because of unmet customers needs, and technology push ideas, those which are generated from the drive to exploit an existing technology by finding additional uses for it.[121] Whitmore (1997; pp. 136–137) emphasizes that "a strategic fusion of market pull and technology push is the answer" to the question of how innovations will be generated in the future.

Table 13 outlines some of the differences between the existing toolkits, the lead-user method and the concept of the license to innovate. While in the license to innovate users are in some way integrated into a companies' routine R&D process, the ties are much looser in the other two methods. Design toolkits are not intended to generate innovations but are often merely used for profitability reasons (Schreier 2005). While the lead-user method is often an excellent way to obtain improvement ideas and sometimes even implementations of these (Herstatt 2004), it has however become a routine part of very few companies' R&D (cf. Olson and Bakke 2001). It may even lead to a prisoners' dilemma as especially corporate lead-users are often, especially when standards are still being established, reluctant to disclose vital information unless they can participate in the IPR.[122] As this can adversely affect their marketing and technological freedom, manufacturers have to be careful with such arrangements. In dynamic markets, where users may one day end up being a competitor of the initiating company, there seems to be an incentive problem for lead-users to disclose crucial know-how or market trend information. While Herstatt (2004; p. 22) found that individual lead-users are frequently willing to allow the initiators of the project to obtain the IPRs without any significant compensation, he mentions that there may be a substantial problem if users have already performed significant work on the project, such as building a prototype. In such cases they may be reluctant to cooperate without appropriate compensation. Through licenses these problems may be largely solved as clear-cut contractual solutions can clarify the ownership of any IPRs that may be

*Table 13* Design-Toolkits vs. the Lead-User Method vs. the License to Innovate

| | *Toolkits for User-Design* | *Lead-User Method* | *Licenses to Innovate (Toolkits for User-Innovation)* |
|---|---|---|---|
| **Focus** | User-customization/ design | To obtain innovative ideas and solutions to problems | Product improvement |
| **Involved Users** | All | Select users | Select users |
| **Required User-skill level** | Low | High | High |
| **Effort required of users** | Low | Low/medium | High/considerable |
| **Scope of Design Space** | Limited/constrained | Open | Wider & more open |
| **Contractual Ties** | Low (potentially through sales contracts) | Low (e.g. confidentiality agreements) | Strong |
| **Amount of explicit know-how that is included** | None or very little | None or very little | Potentially substantial amounts |
| **Included equipment** | Software | Generally none | Software and Hardware |
| **Legal Risk for Tinkering Users** | Medium—no contractual right to tinker; threat of patent infringement or EULA violation | Medium—no contractual right to tinker; threat of patent infringement or EULA violation | Low—Regulated by license; although possible liability for breach of contract |
| **Integration into companies Routine R&D process** | None or very low | Low | Medium |

Source: Own Table

developed ex-ante. As licensed users can be given a contractual incentive (financial or in some non-pecuniary way) to improve the devices in question and disclose the findings of their tinkering, both parties can benefit.

Such a contractual tie allows the user to play a more integrated part of the companies' R&D process. A well designed license contract could lead to a closer and longer-term relationship than under the lead-user method. In this manner an increased amount of trust could be build up to overcome the potential moral hazard problem of each party not actually transferring its best know-how due to the fear of misappropriation and non-reciprocation.[123] We believe that a main advantage of granting a user a license to innovate, which are granted in respect of certain predefined products, is the possibility of tailoring the user's improvements to the competences of the manufacturer. If the computerized dentistry manufacturer would grant its users a license to improve the software by for instance revealing source-code and granting specific programming tools to select lead-users, any resulting improvements could probably easily be implemented by the manufacturer. With respect of the general lead-user method there is no assurance that the resulting suggestions are tailored to the manufacturer in question. The best ideas suggested by lead-users may not be the best solution for the respective company. Financial restraints, the lack of appropriate know-how or skilled employees or merely the existing corporate strategy may be reasons not to pursue the lead-user solutions. Consider a small medical device company that manufactures implants. In its quest for new possibilities for innovations some leading physicians for instance report that they are increasingly operating using regenerable materials. While this may represent an excellent solution for the patient it may not be desirable for the company in question. It may for example believe that pursuing such radical innovation would hurt its existing sales or that that it does not possess the competences (know-how, resources, staff, instruments, etc.) to implement such solution. Such idea would therefore most probably be discarded. While we have no empirical data as to how frequently lead-user ideas are found not to be useful to the respective company, we have some indications for this proposition from a major employer of this method. This issue seems especially acute for small, highly specialized companies.

Due to the potential inclusion of equipment and know-how licenses to innovate can allow users to become a form of R&D facility, which are especially useful for product testing. We argue that such licensing can allow the parties to jointly overcome difficult situations of "high functional uncertainty" coupled "with very sophisticated technical problems" (Lotz 1992; p. 59). Douthwaite (2002) emphasizes that through such close co-operation with users these can "finish" a technology, i.e. implement the technology in the organisational routines by learning from use-experience.

Such licenses could also be seen as a product-service systems (PSS), namely combinations of products and services that are tailored to a specific need of the user.[124] The service would not only be to teach the user how to utilize the product, but also to help him modify it. One example of a PSS may be to grant a doctor a toolkit to allow him to tinker with a certain machine. In other words, granting the user a specific solution space, including certain tools, in which he can freely tinker without thinking of the legal consequences. In this respect "developers ha[ve] to be willing to play the role of tutors rather than [merely] providers" (Leonard 1998; p. 102). While many companies help their customers to use their products effectively, such as via help-lines, few teach them how to modify. Consider the case CACTUS, a proprietary software program for creating country-specific operating interfaces for the manufacture of equipment. In CACTUS, the developers transferred the software to Japanese users and the later were subsequently highly dissatisfied with the education they received as they learned how to "use CACTUS—but not how to change it" (Leonard 1998; pp. 102–103). As an illustration, Sony or Lego could teach its users how to develop software applications for the Aibo or the Mindstorm.

Currently the use of licenses to innovate for users is rather limited as far as we are aware. GE for instance recognized that in the medical imaging business many of the major clinical software innovations were developed by users. Consequently, they selectively supplied certain lead-users likely to develop innovations with very low priced machines in return for which these contractually agreed to give GE preferred access to any improvements that they may develop. This practice has provided GE with various commercializable improvements.[125]

Intel also donates a large amount of equipment to universities and even trains faculty members how to use these. Intel then monitors the use of such equipment. In case any improvement is generated from it, Intel has secured itself not only an automatic contractual right of information, but also a royalty-free right of use.[126] Under such particular MTAs users benefit substantially from the possibility of an economic acquisition of new devices, the possibility of obtaining complementary know-how, access to trained staff, an opportunity to gather early usage experience, publications, seminars and increased prestige, especially if their tinkering leads to improvements. For the industrial partner it offers the chance to collect user-oriented improvement suggestions even before wide commercialization, to stay at the leading edge of the industry, to benefit from the signaling effect of the publications and seminars as well as from a certain empathy of the user who is embedded into the R&D process.[127] Another example of enabling users to improve its technology is Philips Electronics' Pronto universal remote control. Phillips has granted a group of hackers access to program files, codes and arranged for other manufacturers of related audio-video products to also post their code to encourage users to program.[128] Although there was no license in this particular agreement, this type of arrangement has benefited both the manufacturer and users, especially as it allows the avoidance of the costs of reverse engineering.[129]

It has to be understood that neither design- nor innovation-toolkits make strategic sense in every realm. As Chesbrough and Kardon Crowther (2006; p. 235) emphasize in respect of the overall role of external parties for internal R&D: "Leveraging external research may function more as a complement than a substitute in the performance of internal R&D activities." In very homogenous markets, it may make very little economic sense for manufacturers to offer toolkits or licenses. In markets characterized by a high heterogeneity of user demand however toolkits and licenses to innovate can make a lot of sense.[130] Additionally, areas characterized by a high amount of information-goods seem much more suitable for such licenses than areas where there are a large amount of physical components. The multitude of existing toolkits in the software industry provides some support for this view. By granting different kinds of toolkits to different types of users a market segmentation strategy can be organized, giving different types of rights to different types of users. Interestingly, Prügl and Schreier (2006; p. 247) explain that, at least with regard to the *Sims,* the use of different tools can lead to the occurrence of different types of innovative activities. Only the most sophisticated and dedicated users would merit the granting of the most sophisticated innovation toolkits.

To summarize, the license to innovate can be seen as a useful instrument to integrate users into the R&D processes of a company. Such instruments can not only incentivise but also enhance the ability of users to innovate, while at the same time enabling the toolkit provider to benefit from this activity. In order for such transactions however to create win-win situations for both parties toolkits have to be applied in the right manner and in the right situations.

## CASE STUDY IN THE PLANT BREEDING INDUSTRY

After having seen that a permission culture can be said to characterize innovation in various realms, let us now consider how such a concept of selective innovation works in the area of plant breeding. This realm offers an interesting field of study in order to demonstrate that licenses to innovate can even function in areas of low user-innovation. While the few improved varieties that international plant breeders have created have led to tremendous yield increases over the last decades many problems remain to be solved. While such varieties were excellently suited for ideal soil conditions and irrigation systems, in unfavorable ones, such as in Africa or the Middle East, their expected benefits failed to materialize. The agro-ecological as well as the socio-cultural conditions were simply too heterogeneous for the standardized seeds to function adequately.[131] As indicated by the user-innovation literature in case of highly heterogeneous user needs, users themselves are often the most appropriate innovators. The best functioning seeds in Africa are indeed still often of local origin.[132] As the African yields have however stagnated at insufficiently low levels and given that further strong yield increases are needed globally to feed the 8 billion people that are expected to populate the world by 2020, collaboration between farmers and seed-breeders is urgently required.[133] Only the combination of a high awareness of local ecological conditions and farming practices possessed by the farmers and the technological acumen of plant breeding scientists can resolve the dooming food crisis.[134]

Such farmer-plantbreeder collaboration could take the form of granting farmers a type of a toolkit to innovate. Farmers are not only provided with varied genetic materials to test and appraise and with various mini-kits of seeds and fertilizers to experiment with, but also receive considerable technical know-how from the scientists.[135] In this way, the scientific expertise, such as in terms of analyzing diseases and developing resistant high-yielding crops, can be combined with farmer's ecological know-how to yield optimal results for the particular environmental conditions.[136]

Before further discussing this however, we have to remember that plant breeding in developing countries is very different than in developed ones. In these regions the principal concern is not the commercial mission to maximise profits but instead a developmental one of helping the world's poorest out of their misery. As Vernooy (2003) reemphasizes, we are concerned not only with increases in productivity, but also equally if not more importantly with diversity enhancement and empowerment of the poorest farmers.

While for many years scientists did not recognize the potential contributions that farmers could make to the seed innovation process, since the early 1980s there seems to have been a gradual recognition of farmer skills of selecting seeds to suit heterogeneous local conditions.[137] Chambers (1983; pp. 92–93) referred to peasant farmers as "the single largest resource not yet mobilized in the development enterprise." Various studies have already reported successful collaborations between farmers and scientists. Biggs and Clay (1981) for example discuss the successful development of dwarf-yielding wheat varieties in Mexico that resulted from the close contact of plant breeders with Mexican farmers. Hossain and Islam (1985) found that rice yields increased in those regions where both farmers and scientists effectively collaborated. Maurya (1989; p. 13) also found that

> by 1987, a total of nine distinctly promising lines had been identified by [Indian] farmers. This had been achieved within a much shorter period than normally necessary to produce only one official release and whereas an official release might prove unacceptable when transferred to our resource-poor target group, those breeding lines which they have themselves selected under their own physical and management conditions are by definition appropriate and acceptable to them.

Chambers et al. (1989; p. 182) therefore prescribed a model of "farmer first," where the "main objective is not to transfer known technology, but to empower farmers to learn, adapt, and do better." In other words the aim is to grant them a toolkit which can enhance their own experimentations. Two principal methods of integrating farmers into developing improved varieties have evolved over the years: Participatory plant breeding (PBB) and participatory varietal selection (PVS).[138] PBB involves encouraging farmers to resume breeding crop varieties in their own fields with scientists adding their skills and knowledge resources. In this process, in which farmers are involved in the whole cycle of R&D activities from selecting experimental material to evaluating these materials to releasing completed material and finally to ensuring that it is adequately distributed, the farmers are treated as partners rather than patients.[139]

While PBB is however still the exception rather than the rule, PVS is far more common.[140] PVS involves offering farmers finished or nearly finished cultivars to evaluate and select the appropriate ones for themselves by growing these in their fields. Participatory plant breeding (PPB), the term for such farmer-scientist collaborations, should be seen as a continuum with various different forms of interaction.

Both PBB and PVS have been reported to have resulted in successful varieties. Cultivars selected by the farmers under PVS programmes have for example been found to perform considerably better in the relevant environments than the ones chosen for general release by plant breeders.[141] Nevertheless, with PVS not only is the contribution of farmers clearly limited to the adaptation stage, but also PVS trials are often managed under a recommended package of practices that is determined by scientists.[142] As PVS has been found to be inappropriate when conventional breeding has not been able to produce cultivars that were appropriate for local conditions, PPB was found to be necessary in some areas to overcome the idiosyncrasy of the local conditions. Vernooy (2003; pp. 56 and 65) calls the results of PBB in recent years "impressive" and reports that in various regions of the world from Nepal to Cuba to China, new varieties have been developed via PPB. He explains that these "experiences make it equally clear that farmers can and should play a key role in the research process through PPB" and concludes by arguing that such "decentralization should replace centralization as the main organizing principle to address specific [heterogeneous] local contexts."

One of the best examples of PPB is MASIPAG, which if translated means "Farmer-Scientist Partnership for Agricultural Development." Formed in 1985 in Los Baños in the Philippines, MASIPAG is a co-operation between farmers and scientists where the initial goal was to teach farmers how to hybridize rice.[143] Using simple tools such as sharp scissors, tweezers, paper bags and cardboard farmers cut of the male parts of the rice flower to stop self-pollination, take pollen from the other parent and sprinkle it on the unsnapped ovary. Through such process of sexual hybridization HYVs have been crossed with traditional varieties to create popular, disease-resistant plants with relatively high yields.[144] MASIPAG has enabled, within certain obvious limits, farmers to breed plants 'scientifically'.[145] Clearly, such hybrids are relatively basic and yield considerable less than professionally bred varieties given the advancement of hybridization research.[146] Furthermore, while PPB is relatively easy with self-pollinating crops such as rice, it is much more demanding with cross-pollinating crops like maize as this would involve arranging an isolation distance and controlling pollination.[147]

While it seems tempting to grant such "toolkits" to the whole of the developing world, resource restraints make such undertaking impossible. Bellon and Morris (2002; pp. 13–14), researchers of the Consultative Group on International Agricultural Research (CGIAR), therefore explain that while PPB methods are well suited working with a small numbers of

*Table 14* Different Breeding Models

| | *Traditional Farmer* | | *Complete Participatory (Full PBB)* | | *Efficient Participatory* | | *PVS* | | *Scientific/ Modern* | |
|---|---|---|---|---|---|---|---|---|---|---|
| *Activity* | *F* | *S* | *F* | *S* | *F* | *S* | *F* | *S* | *F* | *S* |
| 1. Selection of source germplasm | ■ | | ■ | ■ | ■ | ■ | | ■ | | ■ |
| 2. Trait development | ■ | | ■ | ■ | | ■ | | ■ | | ■ |
| 3. Cultivar development | ■ | | ■ | ■ | | ■ | | ■ | | ■ |
| 4. Varietal evaluation | ■ | | ■ | ■ | ■ | ■ | ■ | ■ | | ■ |

Source: Bellon and Morris (2002)

■ = Party engages in designated activity (Farmer=F; Scientist=S)

farmers there are some indications that it would be "very expensive to work with hundreds of thousands of farmers scattered across an enormous area." This seems to point towards selective collaboration with users, even without commercial objectives and the considerations for the protection of IP. Even ICARDA, the International Center for Agricultural Research in the Dry Areas, which had successfully employed the PPB method, had at the time only collaborated with around 50 villages and a total of around a 100 farmers.[148] Witcombe (2005) further added that as he found no example where farmers actually were granted all the instruments and breeding material to take over the whole breeding process, that in all existing PPB methods, plant breeders were still in control, as diagram 13 illustrates.

Due to the research intensity of PPB for both farmers and researchers—it takes around 5–8 years of scientist-user interaction to leave behind "farmers with an enhanced capacity to innovate" (Bunch 1989; p. 56)—Douthwaite (2002) proposed that it should only be used if PVS fails to yield adequate results. PVS only requires a small amount of additional resources and has been a considerable success in many marginal areas.[149] Another issue with PBB is that various national authorities have not been willing to consider data generated using participatory methods when they evaluate varieties for registration and release.[150] Existing IPR systems furthermore are not compatible with the recognition of collective rights.[151] Such legal barriers to PBB could potentially be overcome by a clear allocation of IPR rights, such as via a copyleft license or the creation of a public domain IPR holding agency.

Where does this discussion lead us? We have seen that in very heterogeneous conditions the contribution of farmers in the innovation process can be crucial. By granting farmers "toolkits," i.e. the ability to enhance their experimentation, their previously inadequate skill-level can be raised. Improved varieties for idiosyncratic ecological conditions can result through such collaboration. Are such collaborations useful in the developed world? The example of the Dutch potato breeding company Meijer indicates that they can be. Meijer sends a couple thousand first year clones out of various crossings to preselected farmers. These grow and evaluate the clones, selecting the best ones over a period of 3 generations. Meijer subsequently tests whether such selected clones could be turned into a commercializable variety. Co-operation with farmer breeders increases the chances of finding promising clones as the farmers can screen and scrutinize a large number of clones every year. The farmer usually gets the clones for free and is not paid for his selection work. Only when one of the clones ends up as a variety is he paid on a 50–50% basis per area of seed potato produced by that variety.[152] In this way the skills of the breeder and the farmer can complement one another to lead to a wider variety of successful potato varieties.[153]

The norm however in developed countries is that farmers are largely excluded from the innovation process. Whether this will change depends on a strategic evaluation by plant-breeders. While it may occasionally make sense to integrate select farmers into the breeding and testing activities in the vast majority of cases however, especially when geological conditions are heterogeneous, trait improvement and cultivar development "can be carried out very efficiently by station-based plant breeders . . . It is difficult to imagine how involving farmers in these activities will lead to improvements in breeding efficiency" (Bellon and Morris 2002; p. 20).

## OUTLOOK FOR FURTHER RESEARCH

This work was motivated by the high potential of users to participate and reshape the innovation process. While a large part of this manuscript took a sceptical perspective on the progression of this phenomenon, this was intended as a response to the previously existing literature that was perhaps sometimes excessively one-sided. While much of this study was explorative and many of our assertions require further empirical testing, it should serve as a complement to the current debate on the sources of innovation. As time proceeds, the heights of cumulative research will continue to rise and the scholars and practitioners of tomorrow, will decide which research will be allowed to contribute to shape the information commons in this field and which will be discarded into black oblivion.

To summarize the main conclusions of this present research, we were the first to identify and describe the principal barriers to user-innovation. Legal, economic, technological and social barriers where scrutinized in depth. While we tried to keep our analysis as general as possible, the discrepancies between different realms were found to be considerable. In order to obtain a deeper understanding of the operation of barriers to user-innovation in a specific area we found empirical proof for the existence of such barriers in the realm of computerized dentistry. We determined that numerous hurdles for innovative user-tinkering existed in this area. While software- and hardware-complexity were identified as the most critical barriers, all enquired factors were held to be at least moderate obstacles. A total of 956 responses to the question, or 52.8%, claimed the barriers to be very high. Less than 1/3 of all respondents classified the barriers as non-existent or low. Together with the low degree of user-modifications that we found this demonstrates that legal, economic, technological and social hurdles can restrain user-innovation in the same way as they could potentially encourage it.

Further testing of these barriers is both necessary and desirable. We assert that the dynamic evolution of the individual barriers and their interaction with one another provides a highly fruitful field of study. In the seed-innovation study we saw how barriers can change over time to facilitate or burden user-innovation. We witnessed the evolution of a user-driven field where for a long period of time no manufacturers existed and users innovated in harmony with nature. After Mendel's laws on genetics were fully understood however the practice of professional plant-breeding evolved and the superiority of professional scientific plant-breeders, especially in homogenous conditions, soon became evident. The split between farmers and plant-breeders became increasingly evident and the agricultural industry today is characterized by a very low degree of user-innovation, mainly as third world farming is more heterogeneous and its industrialization has so far proceeded very slowly. We saw a similar trend in other areas such as sport equipment and hypothesized that increasing industrialization/professionalization often had the effect of creating a separation of labor between users, especially if these are non-professionals, and manufacturers. This indicates that the barriers to user-innovation and therefore the functional sources of innovation change dynamically over time.

We strongly believe that without closely scrutinizing the involved inherent barriers the industrial and temporal differences in user-innovation cannot fully be understood. We therefore stress that understanding the prevalence of user-innovation requires a consideration of all the surrounding and influencing factors. We recommend that the barriers to user-innovation should be studied longitudinally by collecting data about an industry over longer periods of time to adequately portray the evolvement of the functional sources of innovation. While such collection of data may be cumbersome we believe that it is the most appropriate. Although in our seed-innovation case study we scrutinized the evolution of the industry over time, we obviously lacked precise data about the functional sources of innovation. While in this scenario the rise and fall of user-innovation were very clear, in other realms this may not be the case and historical data collection may be far more important.

The development of open-source communities, which have blossomed in a wide variety of areas from software to biotechnology to cars, currently provides a rich field of study. While the ideas that such users have, such as creating open-source cures for Tropical diseases and constructing open-source cars,[154] seem fantastic their practical feasibility is likely to depend on outside support. Scrutinizing how such communities operate and what factors constrain their work could give an important insight into how such barriers could be overcome and how user-innovation can be encouraged.

In this sense the collaboration between proprietary and open-source models seems fascinating. Time will tell what respective roles the two distinct forms of creation will play. What seems certain at this stage is that the forthcoming years will be increasingly characterized by the interaction of the two. Users are likely to play a critical part in shaping the innovation processes of tomorrow. Modern technology has empowered the individual which has lead to the intriguing democratization of user-innovation argument. This holds that users, both as firms and individual consumers, are increasingly able to innovate by themselves. This wonderful vision seems to be turning into reality in the form of the open-source software movement. Nevertheless, the democratization perspective does not offer a complete view of reality. We have therefore attempted to enrich the literature in this area.

We have argued that free omnipresent user-innovation is not in everyone's interest. Innovation can be a profitable endeavor which the incumbent players are not likely to give up to users for charitable reasons. While in some cases it will make strategic sense to allow, encourage and even support user-innovation, this will not always be the case. We have seen how the progression of technological complexity, anti-tinkering methods, the rise in importance of IPRs or frequent restrictive market practices can burden the free tinkering of users. We saw how incumbent players can purposely impose barriers to make user-tinkering more difficult. Given the deeply-entrenched roots of the Confucian philosophy of learning by imitating the exemplatory in the Chinese economy,[155] which is likely to increasingly influence global economic affairs, the resulting wide-spread ignorance of IPRs leading to a quasi-epidemic copying of Western products, seems likely to further increase the incentives to create products which are difficult to copy, modify and reengineer.[156] Ordinary users will frequently not be able to overcome such barriers. The distinction between individual amateur users, individual professional users and corporate users is thus critical. While some users or user-communities are still likely to develop innovative products, in the areas that we considered from seed-innovation to medical equipment or sports equipment, we did not find many indications for a democratization of user-innovation. Instead we found numerous signs that we may be seeing some form of professionalization of user-innovation. While in regard to information goods such possibility certainly exists, in regard to non-information goods there seems to be very little support for such assertion. Future research therefore needs to clarify whether user-innovation has become easier or more difficult in various realms. Both perspectives merit further consideration and evaluation.

While the permission-culture that we discussed may not be a comprehensive characterization of reality, its possible effects should not be ignored. We consequently argued that the logical consequence of such permission culture is that licensing will become increasingly prevalent. Regardless of the evolution of such barriers we believe that licenses are and should be considered to be an important instrument to generate innovations. Unlike mere users licensees have both the legal permission and, due to the know-how or tools that they obtained from the licensor, also an enhanced technological ability that may enable them to generate improvements. They therefore can receive considerable help in overcoming the discussed barriers to user-innovation. We hope that we have clarified the potential of licensees to improve upon technology in tremendously influential ways.

We therefore argue that it can often make sense to integrate select users into the innovation generation process. How to design license contracts so as to both enable and encourage users to improve specific technology, while granting some control over such developments to the manufacturer, will be an exciting question for future research. A way has to be found to practically combine the potential of both users and licenses as an alternative method of user-manufacturer collaboration. Such collaborations to embrace user-innovation could be welfare enhancing as the study by Henkel and von Hippel (2005) indicates.

Work on how to choose appropriate users with whom to collaborate also seems promising. The current approaches need to be adequately tested and improved upon. The strategic question of how many of such licenses should be granted is worthy of consideration. Should every user be able to receive such advanced toolkit or only those who seem most likely to be able to improve the technology? While involving more users increases the probability of a valuable improvement being generated by these, due to the wider dissemination of important know-how it also increases the risk of appropriation by competitors. It therefore seems likely that a wide set of different relationships will exist between manufacturers and their users. Different users should be empowered in different ways. There should be numerous different types of toolkits. Whether the best method of choosing who should receive which toolkit is merely the market via price segmentation, charging more for more advanced toolkits, is highly debatable. Should users pay for these toolkits in the first place? Whether systematically searching for licensees or allowing these to self-select themselves to qualify for such licenses is another unanswered issue. The likely responses to these questions will depend on a strategic evaluation of the specific technology, parties, markets and environments concerned.

Whereas Prahalad and Ramaswamy (2004; p. 238) conclude by saying "welcome to the future of competition and the opportunity to co-create that future," we would like to add that co-creation will have its clear limits. We doubt that a universally applicable paradigm shift in favor of user-innovation has occurred. The ability to control the way in which a product is used for instance may even increase in the future, arguably decreasing the bargaining power of the user. The future of co-creation depends on how the global market place and its regulators embrace such concept. While the open-innovation paradigm seems inescapable given how R&D costs and technological complexity have evolved, we believe that the transfer of a part of the innovation-task to users is not an inevitable outcome. Before we will experience a democratization of user-innovation the existing barriers to user-innovation have to be fully understood and ways have to be found to overcome them. Controlled, rather than a free and democratized, co-creation of users may therefore be a more appropriate characterization of both the current and future status quo. Akin to the political sphere, where in theory the ultimate power also lies in the hands of the electorate, researchers and practitioners should be prudent to analyze where the actual functional locus of innovation resides. If the people and national authorities really want a democratization of innovation, they have to fight for it. Given the expected tremendous technological advances in the forthcoming years and their effect on the human species itself, broadening the innovation base is indeed a desirable and extraordinarily important pursuit for the societies of tomorrow. We hope that we participated in this combat by identifying and exploring some of the enemies in question.

# Appendix
## The Lead-User Method in the Medical Device Industry

### INTRODUCTION

In order to ensure their long-term growth, companies require a balanced portfolio of both incremental and radical innovations. In the ideal situation, the continuous improvement of the existing product/service range would provide the financial support for the ongoing development of the radical innovations. In successful companies, the development of incremental innovation projects belongs to the day-to-day routine work. Management possesses an arsenal of methodologies, with which these projects can be systematically planned, steered and controlled. Traditional market research therefore offers many possibilities to determine the needs of the customer and to test the developed concepts in the target markets prior to market launch.

However, it is a very different situation for so-called breakthrough innovation projects. In such cases, the results from market research studies continually evoke disappointment, due to the difficulty of determining the demand for tomorrow's markets using classical need-assessment methods. This may be due either to the failure of traditional methods to enable customers to articulate innovative ideas or to the fact that the typical customer involved in market research studies is generally unable to look beyond the current product generation.

Leading companies such as 3M, HILTI, Nortel Networks or Kelloggs are increasingly working with lead-users in the early phases of innovation projects (Lüthje and Herstatt 2004). Lead-users are especially well-qualified, advanced users who are both sufficiently well motivated and qualified to make significant contributions to the development of new products or services (von Hippel 1988). Their identification and involvement is supported by the lead-user method as originally described by MIT Professor Eric von Hippel and since employed and improved upon through numerous studies by von Hippel and his scholars (www.lead-user.com).

## 1. WHAT ARE LEAD-USERS?

Lead-users are identified predominately through two characteristics:

1. *Lead-users anticipate the future needs of the market and do so significantly earlier than the majority of other customers.*
2. *Lead-users profit strongly from innovations that solve their problems or enable new opportunities.*

The first characteristic is necessary to determine whether a user possesses the potential to make valuable contributions to the development of new products. Through their prominent role in the recognition of new challenges and application problems and their strong dissatisfaction with current market offers, they can serve manufacturers as predictors of future market needs.

The second lead-user characteristic is about the intrinsic motivation for innovation. The expected benefits associated with an innovation, especially if the current market offering fails to fulfil their needs, can become so strong that lead-users become innovators themselves.

## 2. TARGETING LEAD-USERS

In the 1980's von Hippel developed a systematic approach to search for lead-users. The so-called lead-user method has since, on the basis of numerous applications, been further improved upon. The methodology is based on a four-phase process that begins with the delineation of a search field and ends with the development of a product concept. Applications of the methodology by leading manufacturers have so far lasted between 4 to 9 months. Diagram 1 above summarizes the various steps of the lead-user approach.

### The Lead-user Process

#### *Phase I: Start of the lead-user Project*

A lead-user project is too demanding for it to run within a company's functional area alongside routine duties. Hence, the creation of a dedicated, interdisciplinary team consisting of people from marketing, sales, R&D and production is required. The team can encompass 3 to 6 members where a minimum of 50% of their work time should be available for the project. The project team begins with delineation, as accurately as possible, of the search field i.e. a market, product field or service area in which innovative ideas could be found. Goal formulation then follows with the determination of the basic demands required to satisfy the development (desired degree of innovation). Both internal and external influences on the project should be considered at this stage.

### *Phase II: Trend prognosis*

The identification of critical trends and developments in the search field is important in order to confirm that the lead-users detect market needs earlier than other customers. These include technology and market trends relating to each search field as well as the prediction of economic, legal and business developments. Once such trends are known the quest to find lead-users can commence. There are many different sources of information available for trend analysis. In addition to the analysis of secondary sources (e.g. academic publications, data banks, internet), expert presentations have proven to be especially valuable. Numerous experts from a variety of different backgrounds should be chosen in order to ensure that important developments are not missed (e.g. concurrent technologies, newly created markets). Generally even a few expert interviews will grant a good overview of the critical developments in the investigated markets. The interviewed experts often turn out to be a good source for identifying advanced customers.

### *Phase III: Identification of Lead-users*

In order to ascertain who the lead-users are, the project team must first determine the indicators that will allow their correct identification. Amongst other considerations, it is critical that the users actually lead the trends that were chosen as being important in the previous phase. Further indicators that have proven worthwhile in previous applications of the method are the dissatisfaction with current market offerings, a high understanding of problems in the search field in question and the ability to develop their own product improvements or solutions. The process of searching for the lead-users itself is in a creative one that must be tailored to the specific conditions of the relevant search field. The four currently used methods are outlined in part D.2 above.

### *Phase IV: Development of Product Concepts*

Once a group of lead-users has been identified they are invited to an idea generation workshop that typically lasts two to three days. The workshop serves as an efficient discussion platform where the users interact with the company's staff. Depending on the involved area, the issue of IPRs will probably need to be addressed. It is sensible, as a minimum, for the parties involved to sign a confidentiality agreement and to determine the rights of ownership and use. From experience, lead-users are generally prepared to waive the rights to their ideas without any significant form of compensation.

The workshop revolves around the development of problems associated with existing market offerings and the challenges of future solutions. The results of these discussions, aided by creativity and workshop techniques, serve as the basis for the subsequent development of concrete innovation ideas in teams of three to five people. Throughout these processes, the ideas are developed so that by the end they form a basic concept outline or model.

After completion of the workshop, the ideas are subjected to a preliminary evaluation via a presentation to the respective decision makers within the organization. The team members act as product champions during this process. When the decision is made for the further development of a concept, the organization's normal product development and evaluation processes are utilized. The lead-user approach is therefore no substitute for traditional innovation management or market research techniques.

## 3. LEAD-USERS CREATE NEW PRODUCT LINES: JOHNSON & JOHNSON MEDICAL, GERMANY

A recent lead-user project was carried out by the German subsidiary of Johnson & Johnson Medical (referred to as J&J). The management staff of surgical hygiene products, including disposable articles used during surgery (gowns, masks and garments), saw the lead-user approach as a promising tool to develop ideas and concepts for completely new products.

### Phase I: Start of the Lead-user Project

The key decision maker at J&J chose the patient coverings and the protective clothing of operation personnel as the search fields for which innovative ideas should be sought. This product area is in the very mature phase of the product life cycle. There are really no significant differences between competing products causing the customer's choice to be generally based on price only. It is likely that the price competition will increase further due to the rising cost pressures being experienced by hospitals and clinics. J&J therefore was sceptical if the lead-user approach would allow it to generate radically new ideas for this specific product field. The project team consisted of staff from marketing, sales and product development. The senior author of this book served as the facilitator of the process.

### Phase II: Trend prognosis

In order to determine future trends discussions with experts from numerous areas were held:

1. The largest expert group consisted of users from the target market, i.e. of surgeons and leading OP nurses. Those questioned included surgeons who were leaders in their fields and had implemented new technology in their surgery. The group also included 'extreme users', i.e. doctors who worked under particularly difficult conditions such as in the tropics (poor hygiene controls, poor product availability) or surgeons working on burns (long operations with large, moist wounds).
2. In addition to current users, other experts from the target market were involved in the discussions (second expert group). This included hygiene experts, equipment purchasers in clinics and hospital logistics personnel.
3. The third group was made up from experts in analogous application fields, primarily from semi-conductor production. This is as the conditions required in the 'clean rooms' of chip factories are similar to operation rooms (free from dust or particles).

These expert discussions resulted in a large number of ideas for the improvement of the coverings and protective clothing. These largely incremental ideas were considered to be a useful by-product by J&J. The discussions also resulted in a deeper understanding about developments that would play an important role in this search field for the future. The bandwidth ranged from economical trends (the need for the reduction in hospital inventory), to medical developments (the increase of particular infections) to concrete technological trends (new surgical techniques). J&J needed to focus on one idea.

The project team decided to concentrate on the technological trend of surgical robotics. In some surgical fields (like hip and knee prosthetics, minimally invasive heart surgery, brain surgery) surgical robots are standard practice. They support the surgeon in the tasks that require the highest precision possible such as the exact drilling of bones. The increasing use of this technology has created a novel demand for hygiene products that protect against infection during the OP. This was identified as a fruitful field for innovations. For this project, J&J chose to concentrate on hip and knee prosthetics.

Although another J&J-branch creates design components for surgical robotics, the hygienic problems of these had not previously been considered to be an attractive field for innovation by J&J. This was a positive surprise and created confidence that the lead-user approach could be valuable in the chosen field.

## Phase III: Identification of lead-users

Lead-users were sought from surgeons that primarily operate hips. Within this field users were identified who worked at the forefront of technology, identified serious hygienic problems, reported making use of cutting-edge technology and even possibly had made their own attempts to solve these problems. At this point an exhaustive screening process was applied. Through

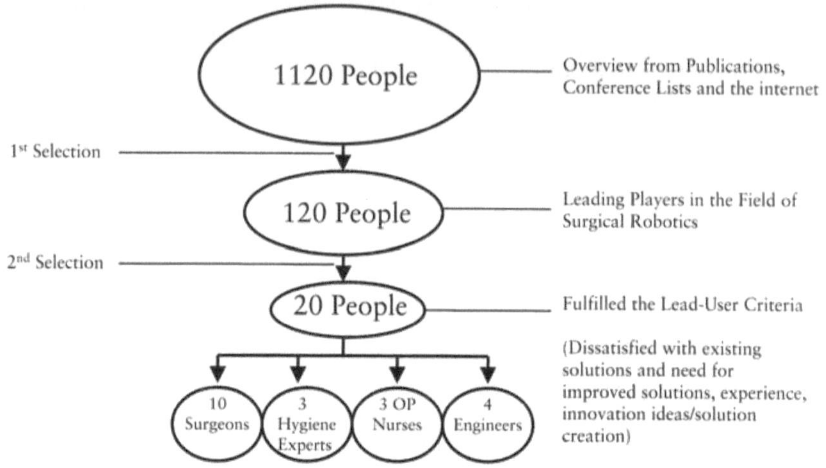


*Diagram 14* Selection process for identifying lead-users.

research of relevant publications and the internet, as well as attending conferences and interviewing experts, around 120 people were identified who play a critical role in robotic surgery today. These people were contacted via telephone or visited in their clinic and questioned. Furthermore, they were asked to name any other people that were leaders and active in an innovative way in robotics or analogue fields. These people were also contacted. The systematic selection process of lead-user is outlined in Diagram 14.

During the selection process, a group of 20 people were identified who matched the lead-user criteria. The members of this group came from four fields displaying varying competencies: surgeons as users of robotic systems, the OP personnel responsible for hygienic conditions (e.g. covering of the robot and patient), the clinical engineers responsible for the technical maintenance and clinical hygiene experts.

## Phase IV: Development of the product concept

From these 20 lead-users, 18 were present for the subsequent 2-day lead-user workshop. This number of participants reflected the high degree of involvement of the group in the development of the product concept. The group was complemented by three robot-manufacturing representatives, three employees from J&J and the senior author of this book. All workshop participants agreed to surrender any rights to J&J.

The group started the workshop with the aim of developing concrete concepts for innovative hygiene products in the field of robot-assisted surgery. Initially the participants involved themselves with hygiene problems in current OP applications with surgical robots and the evaluation of future ones.

This process lasted a half-day and resulted in the workshop's search field being divided systematically into unique and clear sub-fields. From these 4 concrete fields, the participants chose those that needed to be solved with the highest priority. In order to develop these fields further, four sub-groups were formed consisting of 5 to 6 participants and a moderator. In the remaining one and a half days, the group worked intensively on developing innovative solutions and in doing so produced numerous sketches and simple models form a variety of materials. The makeup of the groups varied throughout depending upon the interest of the participants. The work of sub-groups was shared between all groups via presentations and discussions.

At the end of the second workshop day, 4 complete and detailed concepts developed from the numerous individual ideas and solutions. One was the development of a new type of foil to cover the surgical robot that could remove the current sterility and handling problems. The second sub-group suggested a solution for the stifling of the mist that is produced during the operations (so-called aerosols, moisture of water as well as bone and blood particles). The third solution consisted of a new, integrated system for the sterile storage of the patient's leg during a hip replacement surgery. The forth group developed a process for the cleaning of the surgical robot after the completion of the procedure.

In a subsequent evaluation of the developed concepts by the participants of the workshop, the results of the first 3 groups performed outstandingly with respect to the criteria of originality, problem-solving capacity and its ability to be implemented. It should not be forgotten however that the participants of the workshop were made up of leading surgeons, OP nurses, engineers and hygiene experts. The evaluation of the ideas was based on a broad as well as a deep knowledge in the field of robot-assisted surgery.

The workshop provided J&J with strong ideas for potentially new product lines. All 4 of the concepts included products that were not currently in the J&J product program. Some of the product ideas are still not available from any manufacturer in the market, making them world firsts.

## 4. DISCUSSION

The deliberate alignment with the customer and the desire to make compatible, significant and innovative jumps belong to the greatest challenges of innovation management. Companies that are able to do both increase the probability that their new products and services will succeed in the market. As shown here, leading companies can benefit from the intensive cooperation with lead-users.

The effectiveness of the lead-user method has been demonstrated in some first applications. An investigation within 3M, who have great experience with lead-users, supports this. In a comparison between lead-user projects and traditionally carried out development processes, the degree of innovation, turnover, market share, ability to be realized and strategic importance all were measured as being significantly superior in lead-user projects. From these projects, new product lines resulted that without exception were responsible for a large proportion of the respective company's turnover (Lilien et al. 2002, Morrison et al. 2000).

The lead-user application at J&J described above resulted not just in the possibility to further develop their existing market offer but also the chance to expand their program and develop completely new product lines. Despite of all the positive results, it should not be forgotten that the lead-user method is no substitute for proven methods of innovation management and market research. It leads companies primarily to new ways of developing innovative products and services. As a result it produces 'market-fit' ideas for innovations that must be further developed with traditional methods of market research, concept and prototype evaluation as well as product development.

# Notes

## PART A

1. e.g. The introduction of the "Goal Camera" in Hockey or the "Hawk Eye" (cf. Sport1 2007) in Tennis, which both allow the exact verification by camera of whether the puck has crossed the goal line or whether the ball bounced within or outside of the line.
2. See Wagner and McDermott (2006) who discuss the commercialization of Christianity.
3. In the early 1800s for example the Luddites, English textile workers, unsuccessfully protested against the mechanization of their profession by destroying weaving machines. In the long run people's current protests against gene technology are likely to be similarly futile.
4. UNFPA 2006, 2004
5. Jewkes et al. (1969; p. 25) explain that even the US courts have struggled to agree on a single definition of invention and innovation and instead have created "contradictions and confusions."
6. Lüthje and Herstatt 2004, Veryzer 1998a
7. e.g. Lilien et al. 2002
8. von Hippel 1985
9. Herstatt 1991
10. Shaw 1985, von Hippel 1976
11. von Hippel 2005a, Rogers 2003
12. "Advances in information, manufacturing, and basic technology have accelerated so quickly that many processes and products now have lives of three months or less before they become obsolete" (D'Aveni 1994).
13. Chesbrough (2003; p. 35) explains the increasing importance of "learning by hiring away."
14. D'Aveni 1994. "The value of an idea or a technology depends on its business model" (Chesbrough 2003; p. xxx).
15. Gupta and Wilemon 1996. Larry Huston, director of external innovation for Procter & Gamble (P&G), stated "inside P&G are more than 8,600 scientists advancing the industrial knowledge that enable new P&G offerings; outside are 1–5 million. So why try to invent everything internally?" (Chesbrough 2003; p. xxvii).
16. Chesbrough (2003) and Chesbrough and Kardon Crowther (2006) provide some empirical evidence of the open innovation paradigm in both high-tech and established industries.
17. Rice and Rogers 1980
18. von Hippel 1976
19. Lettl et al. 2006

20. Ramírez 1999
21. Kodama 1995
22. e.g. Nohr 2004
23. Lüthje and Herstatt 2004, Morrison et al. 2004
24. Nagel 1993
25. Tuomi 2006
26. Tuomi 2006, Rogers 2003, Kline and Pinch 1994
27. Nuvolari 2004
28. Douthwaite 2002
29. Hienerth 2006, Lüthje et al. 2005, Tietz et al. 2005
30. Lettl et al. 2006, DeMonaco et al. 2005, Lüthje 2003
31. Enos 1962
32. Morrison et al. 2000, Lüthje 1999, Nagel 1993, Herstatt and von Hippel 1992, von Hippel 1976
33. Agarwala-Rogers 1978
34. von Hippel 2005a
35. Mossberger et al. 2003
36. Pearson 2005, Joy 2000
37. Greenberg 2007
38. Tietz et al. 2006, Neff and Stark 2004
39. Feller and Fitzgerald (2002) explain how 3rd and 4th generation programming languages have significantly simplified programming.
40. Cf. Berners-Lee 2000, Prahalad and Ramaswamy 2000
41. Feller and Fitzgerald 2002
42. Cf. Whitmore 1997, Jewkes et al. 1969, plus personal interviews. Nevertheless, a recent MIT Sloan working paper illustrates that clinicians can play a major role in the discovery of alternative uses for existing prescription drugs (DeMonaco et al. 2005).
43. e.g. see von Hippel's (2005) *Democratizing Innovation.*
44. von Hippel 1988

## PART B

1. For a recent example see how mobile phone operators pushed UMTS over Edge solely due to their huge investments into obtaining the requisite licenses (Tiwari und Buse 2007).
2. Freel (2003; p. 763) summarizes that "networking cannot be considered neither a necessary, nor less a sufficient condition for innovation and it may be that, within both the academy and the polity, the tendency to overstate the impact of networks persists."
3. Ledwith and Coughlin 2005
4. Hope 2004, von Hippel and Katz 2002
5. Von Hippel's (1981) argumentation relies on the studies of Taylor and Silberston (1973) and Wilson (1977).
6. Mori 2005, Razgaitis 2004, Chesbrough 2003, Arora et al. 2002, LER 2001. This discrepancy is partly due to the non-consideration of the fact that a majority of all international licensing was due to transaction between affiliated firms that were often largely driven by tax considerations (Contractor 1985).
7. Arora et al. 2003
8. Varian 2002
9. www.freedom-to-tinker.com
10. e.g. von Hippel 2005a, Hope 2004, von Hippel 1988
11. Cf. Schreier 2005

12. Tang 2005
13. Cf. Tuomi 2006
14. IRI 2007
15. Cf. Dahlander 2006, Holt 1988
16. Lilien et al. 2002, Thomke 2002
17. Olson and Baake 2001
18. Klein 1996
19. Peritz 2002
20. Spanbauer 2003
21. See *Arizona Cartridge Remanufacturers Association, Inc. v. Lexmark International Inc., 290 F.Supp. 2d 1034 N.D. Cal. 2003)*
22. Lyra Research 2005
23. According to Black's Law Dictionary this refers to the "process of discovering how an invention works by inspecting and studying it especially by taking it apart in order to learn how it works and copy it and improve it" (Garner 2004; p. 1345).
24. Lockton 2005, Pickler 2002
25. Cf. Maier 2006
26. According to research by the International Data Corporation globally 35% of all software was pirated in 2005, leading to alleged losses of $34 billion (BSA and IDC 2006).
27. Lie 2003
28. Most legal systems contain some sort of defence to a product liability civil action when a product modification occurred. For example "[n]o manufacturer or seller of a product shall be liable for product liability damages where a substantial cause of the injury, death, or damage was a subsequent alteration or modification" [Rhode Island General Laws 9–1–32(b)]. See *Arthur H. La Plante v American Honda Motor Co., Inc., United States Court of Appeals (1st Circuit) Nos. 93–2314; 94–1015*).
29. Rogers 2003
30. Douthwaite 2002
31. The authors experienced that companies may be reluctant to send out questionnaires inquiring about users opinion of the product's quality due to this fear.
32. Focus 2006
33. Dahlander and Magnussen 2005
34. Kline and Pinch 1994
35. This section is based on Bovens 2005, Martinez and Guellec 2004, Kerr et al. 2003, Benkler 2002, Granstrand 1999, Ginarte and Park 1997, Kortum and Lerner 1997, Gana 1995.
36. See *Diamond v. Chakrobarty (1967). 447 US 303, 206 USPQ 193(1980)* for genetic material; *Diamond v Diehr 450 US 175 (1981)* for software; *State Street Bank & Trust Co. v. Signature Financial Group Inc., 149 F.3d 1368 Fed. Cir. (1998)* for business methods. Although in the EU software and business methods are not patentable, thousands of software-related inventions have still been issued (Lloyd 2000).
37. e.g. EU Database Directive 96/9
38. Australia's Digital Agenda Act 2000, the US Digital Millennium Copyright Act and the EU's Copyright Directive.
39. Lerner (2002), analyzing major shifts in patent policy over the last 150 years around the world, found that in the last 20 years there has been a sharp upward trend of strengthening the respective patent system. Cf. Motohashi 2004. Ginarte and Park 1997.
40. In the US, 35 USC 273(b) states that the prior commercial use before the filing of the patent is a valid defence to patent infringement.

41. This is because there are no fair-use or reverse engineering exemptions in patent law (Cohen and Lemley 2001). .
42. Khoury and Sullivan (2005)
43. Harmon (2001) found that the CAFC has shown a great willingness to permit patentees to shut down an infringers business after trial. Recently the Supreme Court in *Ebay v Mercexchange (No. 05–130—15. May 2006)* however emphasized that injunctions should not be granted as a matter of course.
44. 35 USC § 284
45. *Polaroid Corp. v Eastman Kodak Co. 641 F. Supp 828* [228 USPQ 305] (D. Mass. 1985) and 479 US 850 1986) and 16 USPQ2d 1481(October 12, 1990)
46. Rivette and Kline 2000
47. Von Hippel (1988; p. 52) stated that it was "impossible" to invent around "pharmaceutical patents."
48. The fact that various, generally public, institutions have been formed, such as the Fraunhofer TEG or Intracom in Germany, to help companies invent around patents, illustrates the difficulty of such process.
49. Martinez et al. 2004
50. See the 1995 Antitrust Guidelines for the Licensing of Intellectual Property.
51. Jaffe and Lerner (2004)
52. Before the CAFC, the "courts denied 75% of claims against alleged infringers" (Glazier 1997; p. 164).
53. Today, despite the rapid increase in patents, over 60% of patents are being upheld in district court trials (Jaffe and Lerner 2004).
54. To "maintain one's own room to maneuver and at the same time to limit that of the competitors" (Edler et al. 2004; p. 86).
55. Cohen et al. (2002) determined that 81% and 93% of US and Japanese respondents respectively agreed that they use product patents in a strategic way to block other companies from innovating.
56. Hall and Ziedonis 2001, Lanjouw and Lerner 1997. Texas Instrument's and IBM's aggressive enforcement policies have enabled both companies to generate over $1 billion a year (Jaffe and Lerner 2004).
57. US patent litigation increased from less than 1,000 per year in the early 1980s to over 2,600 today (Jaffe and Lerner 2004). This increase is especially substantial as only a margin (less than 5%) of IP infringement cases actually go to court (Lanjow and Schankerman 2004).
58. *Harrington v Yahoo! Inc., No. 4:99 CV-175 E.D. Mo*
59. *St. Clair v. Canon et al.* (2001): Jury ordered that Canon pay over $60 million to patentee (Khoury and Sullivan 2005).
60. *Toshiba v Lexar:* An initial $380 million in damages was awarded against Toshiba (Manners 2005).
61. Hall (2004)
62. Jerome Lemelson, who was granted approximately 558 patents (Varchaver 2001) threatened practically every company in the electronics, semiconductor and auto industries worldwide. Even though it was widely acknowledged that his patents were either unenforceable or invalid, overall nearly 1,000 companies paid Lemelson a total of around $1.5 billion (Shillman 2004).
63. The former FTC chairman Pitofsky (2000) stated that intellectual property "is now a principal, if not the principal, barrier to entry in high tech markets."
64. Somaya and Teece 2000
65. Shapiro 2001, Linden and Somaya 2000
66. Encaoua and Hollander 2002, Eisenberg 1989, Scherer 1970
67. Williams 2005
68. Lüthje et al. 2005

69. James Boyle, American University Law Professor, stated that "at some point, the public domain will be so diminished that future creators will be prevented from creating because they won't be able to afford the raw materials they need" (Shulman 1999; p. 11).
70. Shah (2000) found that 71% of the user-innovators in her study sought to commercialize their products by forming small firms.
71. For instance, Nanotech entrepreneur Markus Krummenacker (2000) pointed out that the "awkward legal language that is used in patents . . . effectively encrypts the patented 'invention' and renders it incomprehensible to the ordinary practitioner in a given field."
72. The American Intellectual Property Law Association estimated that the direct legal cost of an average US patent lawsuit amount to $4 million (AIPLA 2003). Kortum and Lerner (1999) estimated that the cost of US patent litigation amounts to as much as 25% of the aggregate R&D expenditure of US companies.
73. Poynder 2003a
74. Lanjouw and Lerner 1997
75. Lanjouw and Schankerman (2004) estimated that on average it takes between 8 to 25 months to settle an alleged patent infringement.
76. James Watt's absolute refusal to license his steam engine patent stifled innovation in the area (Nuvolari 2004). Cash payments are frequently not accepted (Lanjouw and Schankerman 2004).
77. Intel was accused of refusing to license certain trade secrets to firms that could not enter into cross-licensing agreements (Shapiro 2002).
78. Von Hippel 1988
79. Shah (2000) found that only 17% of user-innovation in the realms of snowboarding, windsurfing and skateboarding were patented.
80. This may be different in regard to semiconductor design firms, which have been termed user-innovators (von Hippel 2005a), and which hold many patents. Chip design innovation is however their precise role in the industry's value chain.
81. Hall 2004, Mahdjoubi 2004, OECD 2004, Kortum and Lerner 1998
82. Telesio 1977. Martinez et al. (2004) found that nearly 90% of respondents considered the effect of their company's patent portfolio on its bargaining power in negotiations to be very or moderately important.
83. Kingston 2000
84. von Hippel 2005b
85. An arrangement in which "two or more patent owners agree to license certain of their patents to one another and/or third parties" (Antitrust Guidelines for the Licensing of Intellectual Property 1995, §5.5).
86. ETC Group (2003)
87. The patent pool was eventually dissolved by the US Federal Trade Commission (*Summit Technology, Inc. v. VISX, Inc., DKT. No. D-92896 (FTC May 27 1999)*.
88. Greenleaf 1961
89. Walsh et al. 2003
90. Infringement may be hard to detect and firms often expect patentees not to sue them due to the involved costs.
91. Berkowitz 2003. In the US there is both an original common law research exemption (*Whittemore v. Cutter, 29 F. Cas. 1120 C.C. D. Mass. 1813*) and a statutory one which provides an infringement defense "for uses reasonably related to the development and submission of information" to the FDA in regard to drug development and commercialization (35 USC §271e).
92. NIH 1998

93. Nanotechnology generally refers to technological processes that take place within an area that is less than 1,000 nanos wide, with a nano being one-billionth of a meter (EPO 2005). The Royal Society and the Royal Academy of Engineering (2004; p. 5) defined nanotechnologies as the "design, characterization, production and application of structures, devices and systems by controlling shape and size at nanometre scale."
94. Lux Research 2007, ETC Group 2005. *The Wall Street Journal* even refers to US Government estimates of the Nanotech markets exceeding $ 1 Trillion by 2015 (Regalado 2004).
95. Bock 2004
96. In the US, nanotech patents have surged over three-fold since 1996 (The Royal Society and the Royal Academy of Engineering 2004). Lemley (2005; p. 601) explains that "universities and companies are rushing to the patent office in record numbers to patent nanotechnology inventions."
97. Miller et al. 2005, Hebert 2005, Regalado 2005
98. For example, patents have been issued on (i) carbon nanotubes [A 2004 review of US patents by Miller et al. (2005; p. 74) revealed that there are over 300 patents in this area: "As nanotechnology continues to develop, the minefields of patents will become more difficult to traverse"]; (ii) semiconducting nanocrystals; (iii) light-emitting nanocrystals; (iv) metal oxide nanorods (v) atomic force microscopes, etc.
99. e.g. AT&T was forced by an antitrust decree to broadly license its basic patents on transistors at low royalties *(United States v AT&T, 552 F. Supp. 131,136 (D.D.C. 1956)*. IBM was forced by an antitrust decree to grant non-exclusive licenses to all its computer equipment patents at reasonable royalties *(United States v IBM Corp., 1956 US Dist. Lexis 3992, at 25–26 S.D.N.Y. 1956)*.
100. Until the early 1980s all software was unpatented.
101. The basic applications of the internet were despite BT's attempts to claim patent protection for hyperlinks, free and not inhibited by patents (Berners-Lee 2000). Lessig (2001) argued that precisely because web pioneers ignored proprietary interests for the common good and relied on open and free protocols did they succeed in creating an innovation commons.
102. Although many individual genomes are being patented many basic biotechnology inventions ended up in the public domain as the basic research on the structure of DNA occurred before universities started to actively patent after the Bayh-Dole Act of 1980 (Smith Hughes 2001).
103. Harvard University owns patent # 5,897,945 on "nano-scale metal oxide nanorods" which was on 33 different chemical elements, i.e. all metals (Maebius 2004).
104. Lemley 2005
105. Huang et al. (2004) in their analysis of USPTO data, found that nanotech related patents fit into 423 of 462 first-level United States Patent Classification categories.
106. Lemley 2005. The flip side of this is that the infringement of nanotechnology patents may be difficult to detect (Miller et al. 2005). This however refers more to being able to avoid the patent thicket at the research stage and not so much at the commercialization stage.
107. Cf. Jaffe and Lerner 2004
108. According to the Nanotech Updates (2004) 13 out of 15 publicly announced nanotech licensing agreements in 2003 were exclusive, i.e. 87%, 4 of which globally exclusive.
109. Ebersole et al. 2005, Walsh et al 2003, Heller and Eisenberg 1998
110. Lessig 2004, 2001, 1999, Litman 2001, Berners-Lee 2000

111. Tuomi 2006, Henkel 2004, Berners-Lee 2000
112. After the 2 Supreme Court cases *Diamond v Diehr 450 US 175 (1981)* and *Diamond v Bradley 450 US 382* 1981). Although in the EU software is not patentable, thousands of software-related patents have still been issued (Lloyd 2000).
113. From 1987 to 1997 software patents as a share of overall US patents increased from 1.7% to 3.8% (Graham and Mowery 2004).
114. Bessen and Hunt 2004. Williams (2005) puts this figure slightly lower at around 16,000.
115. Graham and Mowery 2004
116. Blind et al. 2001
117. Babcock and Foley 2004
118. Tuomi 2006, Raymond 1999
119. The Economist 2006a
120. Microsoft for instance asserted patents on the popular video file format ASF to block distribution of an open-source version that was developed by reverse engineering (Tai 2000).
121. Vaughan-Nichols 2004. Microsoft's Director of Intellectual Property David Kaefer stated that it is not a case of if but when Linux and OSS developers will be forced to license other vendor's IP (Galli 2005).
122. Stallman 2002
123. Ibid.
124. Shulman 1999
125. In Europe continuation patents are not a problem as all patent applications to the EPO are published 18 months after the application (Kaizer and Ronde 2005).
126. The "American Inventor Protection Act" 1999 brought the US into conformity with the WTO requirement that patents are published within 18 months after their submission.
127. As applicants can however "opt out" of the 18 month publication requirement for patents filed solely in the US (35 USC § 122b, 2B), given the size of the US software industry, continuations will continue to be of importance. Graham and Mowery (2004) found that in 1999 continuations still accounted for 21% of all US software patents.
128. *Rambus, Inc. v Infineon Technologies AG 318 F.3d 1081 (CAFC 2003)*
129. Kerr 1996
130. Omura 1997
131. Jackson 2001
132. Babcock and Foley 2004
133. Research institutions and universities frequently ignore existing research tool patents, invoking a research exemption that is broader than the existing legal exemption (Rai 1999). Walsh et al. (2003) explain that there is also a restraint on the aggressive assertion of IP among large semiconductor companies.
134. In January 2005 IBM made 500 of its software patents freely accessible to the OSS community (Galli 2005).
135. US Copyright Law as contained in § 132(a) Title 17 of the United tates Code.
136. "The immediate effect of our copyright law is to secure a fair return for an author's creative labor. But the ultimate aim is, by this incentive, to stimulate artistic creativity for the general public good" (*Twentieth Century Music Corp. v. Aiken, 422 US 151 (1975*). Article I, section 8, clause 8 of the US Constitution provides that Congress shall have the power "to promote the Progress of Science and useful Arts, by securing for limited times to Authors and Inventors the exclusive Right to their respective Writings and Discoveries."
137. e.g. The US No Electronic Theft Act (1997) broadened the scope of copyright protection and for the first time criminalized wilful infringement of

copyright and the reproduction or distribution of copyrighted material by computer users (USC § 506 (a) 2).

138. Lessig 2004
139. Goral 2003
140. Tang 2005
141. For the majority of crimes, deterrence will depend on the certainty of detection and the severity of sanctions (Paternoster 1987, Tittle 1969).
142. 17. USC § 1201(a)(1)(A)
143. §1201(3)(A)
144. § 1203 and 1204. See the *Skylarov* case.
145. Gillespie 2004, e.g. ink-cartridges.
146. §1201(f)(1)
147. Senate Judiciary Committee 1998b
148. See Australia's Digital Agenda Act 2000, and the Amendments to the Japanese Copyright Law and the Japanese Anti-Unfair Competition Law.
149. Directive 2001/29/EC on the Harmonization of certain aspects of Copyright and related rights in the Information Society.
150. See the UK Copyright and Related Rights Regulations 2003.
151. Wragge & Co. 2004
152. Homiller 2005, Dusollier 2003, Hugenholtz 2000
153. e.g. Sec. 1030 US Computer Fraud and Abuse Act, as amended on October 3, 1996. In the UK the Computer Misuse Act 1990 made hacking per se illegal (Sec. 1), providing for both criminal and tortuous liability.
154. e.g. The DMCA or the 2003 US Sentencing Guidelines provide for sanctions of substantial fines or to up to 15 years imprisonment (Hulme 2003a and b).
155. The UK for instance launched a "National Hi Tech Crime Unit" to combat serious cybercrimes. These "digicops" can lawfully intercept information to search for cybercrime (Jarvie 2003a and b).
156. Cf. Hulme 2003b. The US Federal TradeCommission estimated that the damage done due to identity theft alone amounted to over $47 billion during 2002–2003 (Germain 2004). Hackers can also cause immense security risks. For instance in 1997 a juvenile hacker disabled Worcester airport's control tower for six hours resulting in serious threats to public safety (*US v Juvenile (1998) March 18*, USDJ 1998).
157. e.g. Sec. 1 of the English Computer Misuse Act 1990. Samuelson (2001, p. 2028) commented on the DMCA's breadth: "Any data in digital form can be protected by encryption and other technical measures and those who distribute can thus use the DMCA to restrict what scientists and other researchers can do with this data."
158. Krause 2005
159. e.g. in the UK developers of hacking software, and not only perpetrators, have been arrested (Schultz 2002b).
160. EFF 2003
161. Homiller 2005, Library of Congress 2002
162. McCullagh 2002
163. 82 F. Supp 2de 21 [S.D.N.Y. 2000]
164. Lessig 2001
165. Ferguson 2001, Lemos 2001
166. Wible 2003
167. Hantler 2004, Paternoster 1987
168. McGuire 1988
169. Granstrand 1999, Rivette and Kline 2000
170. Brown 2004. Although the high-tech bubble that was forming at the time plays a role in such valuation, it still indicates the importance of intangible property.

171. Douthwaite 2002
172. Peace 2004
173. It makes sense for the US to have a strong IPR enforcement as in the long run manufacturing will happen elsewhere (Khoury and Sullivan 2005). In Japan, despite the considerable strengthening of IPRs in the late 1980s, Hisamitsu Arai (1997), the Commissioner of the Japanese Patent Office, representatively explained that in the 21st century, the only way to adequately exploit the fruits that flow from investments into science and technology, is through even stronger patent rights. Eight percent of the respondents to the survey by Martinez et al. (2004) indicated that the importance of patents would further increase over the next 5 years.
174. Benkler 2002
175. Patents "came to be seen as a necessary part of any competitive academic institution's research policy" (Rai 1999; p. 110).
176. Eisenberg 1996a
177. While IPRs were not essential for the development of various industries, many "new branches would hardly have flourished without the expectancy of acquiring exclusive rights in the results of R&D" (Geiger et al. 2003; pp. 641–642).
178. For example, the productivity of innovation in the pharmaceutical sector has decreased sharply, mainly due to the escalation of research costs and strong generics competition (Piachaud 2004).
179. Granstrand 1999, Arundel 2001
180. e.g. the continuing spread of the illegitimate mass exchanging of copyrighted files (Marsh 2004), now increasingly over mobile networks (Butcher 2004). Microsoft estimates that it loses about $10 billion a year due to pirated desktop software (Glover 2003). The EU estimates that the global trade in pirated goods is worth over 200 billion a year—around 5% of total world trade (Saragossa 2004).
181. Dutfield and Suthersanen (2004) find that there are fundamental problems with most alternatives to the patent system, such as public procurement or the awarding of prizes. Many commentators, such as Lamoreaux and Sokoloff 1997, Scherer 1977 or Taylor and Silbertson 1973 found empirical evidence for the suggestion that weaker patent systems lead to increased corporate reliance on secrecy, leading to a substantial decrease in the dissemination of technology.
182. 17 USC §302a
183. Senate Report 104–315, 1995, Sec. I
184. Lessig 2004, Levy 2002
185. e.g. the Consumer Electronics Manufacturers Association (CEMA); the Software Publishers Association; the American Society of Composers, Authors and Publishers www.ascap.com, with its membership of over 80,000; the Association of American Publishers (AAP); International Intellectual Property Alliance; and PhRMA (Pharmaceutical Research and Manufacturers Association) which in its Mission Statement states that it is a powerful advocate for "strong intellectual property incentives" (PhRMA 2007).
186. Tang 2005, Marsh 2004
187. Samuelson 2003. In 2002 for example they both strongly lobbied for the "Peer-to-Peer Piracy Prevention Act" (H.R. 5211) which would given copyright owners immunity for impairing the unauthorized distribution or display of their works on public P2P networks (Wagner 2003).
188. www.eff.org (accessed February 28, 2008).
189. Under the "Consumer Technology Bill of Rights" users would receive the right to "time-shift" media (recording a TV show and watching it later), the right to "space-shift" media (copying a CD to a portable MP3 player) and the right to make backup copies.
190. e.g. Samuelson 2003

191. While Samuelson (2004) believes that IP arbitrage offers good news for "those concerned that very strong IP rules are impeding innovation," we believe that on the contrary, precisely due to this arbitrage possibility, the pressure to increase global IPRs will remain strong.
192. Sell 2003
193. Ibid.
194. Samuelson 2004, Sell 2003, Shulman 1999
195. Sell 2003
196. Ibid.
197. Hitt 2005. The main problem in China is the lack of adequate enforcement and the poor court system.
198. Mamudi 2005
199. Davidson 2004, Jaffe and Lerner 2004
200. Khoury and Sullivan 2005
201. Recently the United States Supreme Court in *Illinois Tool Works Inc. v. Independent Ink, Inc., 547 US (No. 04–1329, March 1, 2006)* held that the mere existence of a patent does not automatically confer market power upon a patentee.
202. ECIS 2004. EU-trade commissioner, Pascal Lamy, stated that "Piracy and counterfeiting continue to grow every year and have become industries, increasingly run by criminal organisations...[piracy and the violations of IPRs] undermine the survival of EU's most innovative sectors". Between 1998 and 2002 the number of counterfeit goods intercepted at the EU's external borders increased by over 800%.
203. Directive 2004/48/EC On the Enforcement of Intellectual Property Rights
204. Art. 9.1a, 9.1b and Art. 10, 11
205. Although the existing studies indicate that the relationship between the scope of patents and innovation remains unclear (Hall and Ziedonis 2001, Granstrand 1999, Taylor and Silberston 1973, Machlup and Penrose 1950), it is clear that Western innovation markets were "founded upon patents from the earliest date up to the present time" (Report of the 1851 Select Committee of the House of Lords appointed to consider 'An Act to further amend the Law touching Letters Patent for Inventions', House of Commons Sessional Papers, Vol. XVIII (Command Paper NO. 486), p. 281—found in Fisher 2004; p. 15).
206. The Economist 2006d, Lockton 2005
207. e.g. Professional photographer Justin only started to employ independently filled cartridges once the warranty for his printer expired. Previously due to the expense of the device and the warranty he was "deathly afraid of attempting to do anything with the Stylus Photo 1270" (Response to Seattle Times 2006).
208. Percy Schmeiser quoted in Herbst and Volling (2006)
209. *382 F.3d 1374 (2004)*
210. There are 2 main types: shrink-wraps, where the license terms are printed on the packaging, and click-wraps, where the license terms are presented to users before use, with a click being required for affirmative assent.
211. Harper and Udupa 2004
212. Blizzard's EULA in *Davidson & Associates v Internet Gateway* (the "Blizzard case") *30. September 2004, Eastern District of Missouri, Case No. 4:02CV498 CAS*—pp. 7–8). Adobe Reader 2.5 states that users may not "alter or modify the Software or create a new installer for the Software" (Adobe 2007). Microsoft Windows XP Professional End-user license, Oracle's OTN (Technology Network Developer) License (Oracle 2007) and Dell's Software License Agreement contain similar prohibitions (Dell 2007).
213. In the US: *Llan Sys. Inc. v Netscout Serv. Corp. 183 F. Supp. 2d 328, 337 (D. Mass, 2002)*; *Lexmark International, Inc. v. Static Control Components, Inc., 253 F. Supp.2d 943, 949 .D. Ky. 2003),No 03–5400 6th*

*Cir.*; *Arizona Cartridge Remanufacturers Association, Inc. v. Lexmark International Inc., 290 F.Supp. 2d 1034 N.D. Cal. 2003)*; *Davidson & Associates v Internet Gateway* (above). In the EU click-wraps are generally enforceable if the Unfair Terms Directive (93/13/EEC) is adhered to (Maxeiner 2003).

214. Samuelson 2002
215. 17 USC §107. Copying is allowed for "purposes such as criticism, comment, news reporting, teaching, scholarship and research" (*Atari Games Corp. v Nintendo of America Inc., 975 F.2d 832 (Fed. Cir. 1992*) at 843).
216. at 325–326.
217. Wilson 2004
218. Band 2005
219. European Software Directive (91/250) on the Legal Protection of Computer Programs, Arts. 6 and 91
220. Graham and Mowery 2004
221. Cohen and Lemley 2001
222. Shim 2001
223. at 11803.
224. Musker 1998
225. Graaf et al. 2003
226. e.g. Lexmark sells its printer cartridges with shrink-wraps that restrict users to only utilize the cartridge once. Visionary AV Solutions Ltd, a UK Plasma TV and home cinema system supplier, grants purchasers licenses to use the software incorporated in the equipment, which prohibits purchasers from modifying or updating the software without Visionary's prior written consent (Visionary 2006).
227. Woodward 2002
228. Smith 2003
229. *Bonito Boats, Inc. v Thunder Craft Boats, Inc. 489 US 141, at 160 (1989)*
230. An important counter effect however is that some users will switch to OSS. For example, Firefox was downloaded over 100 million times (Montalbano 2005). Although Gartenberg (2005) argues that this was only possible as Internet Explorer was no longer a strategic product for Microsoft.
231. Ring 2005
232. Lyman 2005, McCullagh 2005
233. Spanbauer 2003
234. Siegal and Liu 2005, *Mallinckrodt, Inc. v Medipart, Inc., 976 F.2d 700 (Fed. Cir. 1992)*
235. The Economist 2001a
236. e.g. The top 200 corporations are estimated to control a quarter of the world's productive assets (UNDP 2000 Human Development Report).
237. De Grauwe and Cameron 2003. Apparently, 74% of Americans agreed that big companies have too much influence over government policy and policy makers (Bernstein 2000).
238. Granstrand 1999
239. Gillespie 2004. This is implemented by the DVD Copy Control Association (CSS Specifications, Version 2.6 as of Jan 17, 2005—www.dvdcca.org/css).
240. = Technological foundations on which other products, services and systems are built.
241. =A "set of technical specifications" that provides "a common design for a product and process" (Lemley 2002; p. 1896).
242. Gawer and Cusumano 2002
243. Intel Director, David Johnson, stated that in the semiconductor industry "we are tied to innovations by others to make our innovations work" (Gawer and Cusumano 2002; p. 2).

244. Chesbrough (2003; p. 60) concurs: "Even very good technologies will flounder if they do not connect effectively to outside complementary technologies, while seemingly inferior ones may overtake them if they are better connected."
245. These are common in the telecommunications and computer industries (Lemley 2002).
246. Kaiser and Ronde 2005
247. Gawer and Cusumano 2002
248. Sanchez 2002
249. "Modular architectures are architectures in which component interfaces have been specified to allow the 'substitution' of a range of component variations into the architecture without having to change the designs of other functional components in the architecture," which enables the "mixing and matching" of internally or externally produced components (Sanchez 2002; pp. 668–670).
250. Shah and Tripsas 2004, Baldwin and Clark 2004
251. Wide licensing is often necessary to impose a standard.
252. For instance MPEG, GSM, 3GPP and IEEE 1394 (Tsilas 2005).
253. PCI (Peripheral Component Interconnect) and USB (Universal Serial Bus) (Gawer and Cusumano 2002). Recently, AMD filed an anti-trust suit against Intel for illegitimate market practices that stifled competition in the industry (Laube 2006).
254. Henkel 2004, Anderson 2003a
255. Samuelson and Scotchmer 2002, Utterback 1994
256. Gawer and Cusumano 2002
257. Time Warner, Toshiba, Hitachi, Matsushita, Mitsubishi and Victor Company.
258. Goldstein 2001
259. Walsh et al. 2003, ETC Group 2001
260. Hope 2004
261. The dominant design is the "one that wins allegiance of the marketplace, the one that competitors and innovators must adhere to if they hope to command significant market following" (Utterback 1994; pp. 24–25). Empirical evidence indicates that in most industries, except in some high-tech areas, the dominant design theory holds (Murmann and Frenken 2005, Kodama 1995).
262. Shah and Tripsas 2004
263. Utterback 1994. Henkel and von Hippel (2005; pp. 79–80) further state that users "will in general lack knowledge of the most efficient technical solutions both with respect to the product itself and with respect to processes to manufacture it."
264. Hienerth 2006
265. Switching costs are simply "the costs of changing suppliers" (Von Weizsäcker 1980; p. 1085) Klemperer (1987) identified 3 types of switching cost: a) transaction costs of switching supplier; b) learning costs of getting to know the novel product; c) artificial switching costs created by suppliers, such as compatibility issues, like not being able to transfer files to an alternative system.
266. For example, in minicomputers, hospital CT scanners, telephone PBX switches, high-volume photocopiers and micrographic reproduction equipment (Le 2004, Borenstein 2000).
267. For instance, in *Eastman Kodak Co. v Image Tech. Service, 504 US 451, 455 (1992)* the US Supreme Court held that switching costs and information asymmetries, i.e. the difficulty customers face in obtaining information about aftermarket costs, could operate as a strain on innovation.
268. Despite of the alleged superiority of Linux in terms of price, reliability, security or ability to run on a wider variety of hardware platforms (Gartner Dataquest 2003), over 90% of the world PC's run on Microsoft's Windows (The Economist 2005b).
269. Le 2004, Douthwaite 2002

270. Le 2004
271. Dahlander and McKelvey (2005) found that even among advanced programmers out of 110 only 50 claimed to be Linux users.
272. Wilke and Bandler 2001
273. Shapiro 2000, *United States v Microsoft Corp, July 15,1994, D.D.C. (No. 94–1564).*
274. The European Commission held that Microsoft abused its dominant position by i) preventing interoperability between its own products and those of rivals and by ii) bundling new products into its Windows desktop operating system (*Commission Decision of 24 March 2004 relating to a proceeding under Article 82 of the EC Treaty Case COMP/C-3/37.792 Microsoft*), ordering it to pay a fine of € 497 million.
275. Berners-Lee 2000
276. Gawer and Cusumano 2002
277. *United States of America versus Microsoft Corporation, Civil Action No. 98–1232 (TPJ), "Findings of Fact", United States District Court for the District of Columbia, 5 November 1999*
278. Le 2004
279. Mossberg 2001
280. Cohen 2006
281. Johnson 2001
282. Shapiro 2000, Röller et al. 2000
283. While today the auto and oil industries are spending billions of dollars on fuel cells the low reinvestment rate—BP for instance indicated that they may return up to $65 billion over the next 3 years to shareholders (Financial Times 2006)—shows that much more could have been invested.
284. The Economist 2003
285. Glaser 1999
286. Ellis 2008
287. Jänicke and Jacob 2004
288. Leiby and Rubin 2004
289. President Bush even admitted that the US economy is "addicted to oil" (FAZ 2006a).
290. McNutt and Rodger 2004
291. Utterback 1994
292. Henkel 2004, Lakhani and von Hippel 2003, Feller and Fitzgerald 2002
293. Rivette and Kline 2000
294. Douthwaite 2002, Raymond 1999
295. Nevertheless Firefox 2 contains some functional advancements such as an integrated spell checker for web-based text input forms and safer surfing due to automatic detection of bogus/fraudulent websites (Ogg 2006).
296. Numbers come from the editorial director of SourceForge's parent company Open-Source Technology Group
297. Rammer et al. 2004
298. Cf. Lerner and Tirole 2002
299. e.g. small biotech companies typically have to invest a minimum of $100m before making a return on their investment (Cavalla 1997) and allegedly 5 out 6 commercialized drugs fail to make back their R&D costs (Bodoni 2005).
300. Suarez-Villa 2004
301. Kaiser 2002
302. von Hippel 2005
303. Franke and Piller 2004
304. Teece (1977) found that the cost of information transfer in 26 international technology-transfer projects averaged 19% of total project cost. Mansfield et al. (1982) confirmed these findings.
305. Washburn 2005, Bok 2003, Slaughter and Leslie 1997

306. Horgan 1998
307. Schiffer 2003, Jewkes et al. 1969
308. Renowned economist James Conant stated that "the typical lone inventor of the eighteenth and nineteenth centuries has all but disappeared. In his place in the mid-20th century came the industrial research laboratory and departments of development engineering" (Jewkes et al. 1969; p. 35).
309. Rothwell and Zegveld (1985) explain that in this century there has been a technology explosion with around 90% of our present technical knowledge having been generated since the 1940s.
310. Kortum 1993
311. von Hippel 2005b, Ogawa 1998, Nelson 1996
312. Lüthje et al. 2005
313. Herstatt and von Hippel 1992
314. Franke and Shah 2003
315. Franke and von Hippel 2003
316. Morrison et al. 2000
317. Ashley 1993
318. Maloney 2005, Harish 2004
319. Ashley 1995, 1993
320. www.wilson.com—Dahlin et al. 2004
321. Ashley 1995
322. Braham 1995
323. Through high-tech research by Adidas and Bayer MaterialScience the TeamGeist has become the roundest, the most durable and water-proof ball ever (Baynews 2006).
324. Baldwin et al. 2006
325. Grimm 2004
326. Ibid.
327. While in the US the ratio of R&D costs to sales was 11% in the 1980s, it has increased to over 20% today (Louie 2001).
328. Establishing the safety, purity, potency, and efficacy of a drug to the satisfaction of the FDA is often a long and time-consuming process (Whitmore 1997). Louie (2001) puts the time it takes to go from drug development to market in the 1990s to 14.9 years.
329. Nearly 4 years later progress seems fairly limited. Only work on malaria and schistosomiasis seem to have commenced (www.tropicaldisease.org as of February 2008).
330. Levy (1984) discusses the examples of Sierra On-line and Atarri.
331. Dahlander 2006
332. GNU 2007; Sec. 2b
333. Proprietary software is generally distributed in binary form- consisting of "almost endless lines of ones and zeros" (Dahlander 2006; p. 20).
334. The information is provided on Wikipedia (2007), but could not independently be verified.
335. von Hippel 2005, Meyer and Lopez 1995
336. Henkel and von Hippel 2005
337. Henkel and von Hippel 2005, von Hippel 2005b
338. Graaf et al. 2003
339. Chesbrough 2003, Somaya and Teece 2000
340. Johnson et al. 2004, cf. Tuomi 2006. Jones 2005
341. Cf. Lemley 2000, Allison and Lemley 2001
342. Jewkes et al. 1969
343. Allison and Lemley 2001

344. While Jewkes et al. (1969; p. 90) add that the distinction between individuals and corporations is "unsatisfactory as a record of changes over time," especially due to the general growth in corporations, they acknowledge that there is no alternative method.
345. Chesbrough 2003
346. Kasarda and Rondinelli 1998, Kodama 1995, Hamel and Prahalad 1990
347. Tuomi 2006, Suarez-Villa 2003, Arora et al. 2002
348. It does have to be added that in specialized OSS communities—e.g. in X-ray astronomy and deep space imaging—the inherent complexity poses a considerable barrier as Feller and Fitzgerald (2002) indicate.
349. Thomke and von Hippel 2002, von Hippel 1988
350. Over 50% of the hobbyist patents were in the bottom 10% of the impact range (Dahlin et al. 2004).
351. Akio Morita, former CEO of Sony, stated that "the public does not know what is possible, but we do. So instead of doing a lot of [market] research, we refine our thinking on a product and its use and try to create a market for it by educating and communicating with the public." (Hamel and Prahalad 1994; p. 100)
352. Rothwell and Gardiner 1985
353. Christensen 1997
354. Lettl et al. 2006
355. Prahalad and Ramaswamy 2004
356. Nordmann 2004
357. Baldwin et al. 2006
358. Maurer and Scotchmer 1999
359. Wible 2003, Skibell 2003
360. Rai 1999
361. Samuelson and Scotchmer (2002; p. 1633) stated that "whatever technology can do, another technology can undo."
362. TWST 2004
363. Bruns 2001, Ellenbogen 1997
364. Michaels 2004
365. Mossberg 2001
366. Foster 2002
367. Spanbauer 2003
368. For instance, Intuit's revenues grew nearly 25% after utilizing product activation (Taschek 2003).
369. Microsoft 2006, Galli 2005
370. *United States v Microsoft Corp, July 15, 1994, D.D.C. No. 94–1564, Commission Decision of 24 March 2004, C 2004) 900 final (21 April 2004)*
371. Schultz 2002a
372. For instance, to access Blizzard Entertainment's battle.net online game playing platform you need an access code, written on the purchased CD-ROM, in addition to an encryption algorithm contained in the game software that is needed to communicate—with the server.
373. Granstrand 1999
374. Gillespie 2004, Samuelson and Scotchmer 2002
375. Perry 2005
376. Oakes 1999
377. See Applied Digital Solutions's Verichip, that allows identification and the retrieval of medical information from patients (Bahney 2006).
378. Harris and Yen 2002
379. Fernandez Delicado et al. 2005

380. www.magiqtech.com Swiss. Id Quantique developed a similar application. www.idquantique.com.
381. Stix 2005, Bergstein 2003
382. Hesseldahl 2006, Johnson 2006
383. Graaf et al. 2003
384. Mollick 2004
385. Bolan 2005
386. Ibid.
387. The Economist 2005a
388. Pooley 2001
389. Ostler 2003
390. For instance, Hitachi's tiny "mu-chip" can be embedded in clothing and paper (Prahalad and Ramaswamy 2004).
391. Lie 2003
392. Anderson 2003a, Motorola 1998
393. Conner 2006
394. Varian 2002
395. Conner 2006, Lockton 2005
396. Spring 2003, Wagstaff 2002
397. Lyra Research 2005
398. Anderson 2003b
399. Knapp 2003. Spring 2003
400. Wagstaff (2002) described that printer companies are "building the cartridge cases like Fort Knox, leaving only the brave to fiddle with them."
401. See www.alotofthings.com
402. Anderson 2003b
403. The Economist 2006b
404. Bramwell 2006
405. Computerbase 2005
406. Mello Jr. 2005, Pickler 2002
407. Mello Jr. 2005
408. Pickler 2002, Mello Jr. 2005, www.righttorepair.org
409. Scheeres 2004
410. The Economist 2006c
411. Cf. Potting Solutions 2007, Shapley 1978
412. For instance, the Trusted Computing Platform Group (TCPG) was originally formed by Compaq, Hewlett-Packard, Microsoft, Intel, and IBM in 1999 and now consists out of 200 participating members (Bechtold 2004).
413. This part is based largely on TCG 2007, Dornan 2006, Kuhlmann 2004, Koenig et al. 2004, Bechtold 2004
414. Pollack 2006, Dornan 2006
415. Singer 2005, Merritt 2005
416. Evers 2006
417. Response of the Parliamentary Undersecretary of State Gerd Andres on 26 November 2002 to the question of the Member of the German Bundestag Dr. Martina Krugmann, Bundestags-Drucksache 15/116 on November 29, 2002—found in Bechtold (2004; p. 88).
418. Specifically Philip Lowe—EU Director General for Competition (Lettice 2002).
419. *Macrovision v. 321 Studios, 1:04-cv-00080 (S.D.N.Y.)*
420. Lutzker and Lutzker 2005
421. *Computer and Internet Lawyer* 2004
422. The result would probably have been similar in the EU (Borland 2004b).
423. Löwer 2005
424. Enomoto et al. 2006

425. Zyvex Corporation for example successfully undertook experiments where cubes of different sizes where able to build different architectural structures. Zyvex aims to produce a "Molecular Assembler" which is a "user-controlled fabrication tool capable of creating molecularly precise structures with 3-dimensional capability in an economically viable manner" (Zyvex 2007).
426. Lemley 2005
427. Cf. Schneier 2001
428. Levy 1984
429. See Mulhall (1997a) for various examples.
430. Wellen and Reiter 1998, Felten 2004
431. Mollick 2004, Hulme 2003a
432. A recent survey of UK internet users found that 55% of these illegitimately copied CDs (BBC 2006).
433. A survey of 1,358 adult internet users by the Pew Internet & American Life Project and a study by comScore Media Metrix, which measured actual use of four popular file-sharing programs in the US, both found that the much-publicized lawsuits against illegal file-sharing slowed online piracy (Stevans and Sessions 2005).
434. Shah and Tripsas 2004, Kline and Pinch 1994
435. Paternoster (1987; p. 214) explains that "perceptions of the certainty and severity of punishment do not seem to deter the trivial and infrequent behavior of high school and university students," which is exactly the group which is most active in hacking and downloading (Kremen 1998).
436. Mollick 2004
437. Utterback, 1994
438. Lessig 2004, Wearden 2001
439. Bovens 2005
440. Labrador 2002
441. Bowman 2002
442. Greenpeace 2007, Euler 2006, MASIPAG 2003
443. Douthwaite 2002
444. Lettl et al. 2006, Lüthje 2003, Biemans 1991, Shaw 1985
445. BMBF 2005
446. Lettl et al. 2006, Allison and Lemley 2001, Reiser and Anbar 1984
447. Shaw 1998, Lotz 1992
448. BMBF 2005, Lotz 1992
449. Gelijns and Rosenberg 1995
450. Gelijns and Rosenberg 1995, Biemans 1991
451. Kahn 1991
452. BMBF 2005
453. Spectaris 2004
454. For the US see Termini 2004 or Whitmore 1997. For Europe see the BMBF 2005 or Böckmann and Frankenberger 2002.
455. BMBF 2005
456. Finkelstein et al. 1995
457. BMBF 2005
458. Termini 2004
459. BMBF 2005
460. Lotz 1992
461. Whitmore 1997
462. BMBF 2004
463. *US Supreme Court, No, 95–754 decided June 26*
464. Hantler 2004
465. Hantler 2004, Gelijns and Halm 1991

466. Hantler 2004
467. Shaw 1985
468. Kahn 1991
469. Lotz 1992
470. Weisbrod 1994
471. FDA 1990
472. Finkelstein et al. 1995
473. Blume 1995
474. Gelins and Rosenberg 1995, Hutton 1991
475. Gelijns and Rosenberg 1995
476. Wiedhahn 2006
477. Guinnessy 1997, Bühner 1997
478. Lettl et al. 2006, Lüthje 2003, Lotz 1992, Biemans 1991
479. Interview with the author, March 8, 2006.
480. Wiedhahn 2006
481. Wiedhahn 2006; cf. Sirona Annual Report 2005b
482. Cf. Rice and Rogers 1980
483. Gesellschaft für Zahngesundheit, Funktion and Ästethik
484. Cf. von Hippel 1988
485. Gelijns and Halm 1991
486. Ibid.
487. Cf. Razgaitis 2004
488. If we were not able to disprove the user-modification we included it.
489. Brad Dorsch (abe88) in CDENT Forum (2004).
490. Horst Landenberger in CDENT Forum (2004).
491. Anneken 2005
492. Curran 1984
493. Confirmed by our interviews and various CDENT instruction manuals.
494. Cf. Lessig's (1999) "code" and Reidenberg's (1998) "lex informatica."
495. Cf. Kodama 1995
496. Cf. Leifer et al. 2000, Veryzer 1998a, 1998b
497. Cf. Cooper and Kleinschmidt 1990
498. Chesbrough 2003, Biemans 1991

## PART C

1. Utterback 1994
2. Baldwin et al. 2006
3. Lüthje et al. 2005, Tietz et al. 2005, Shah and Tripsas 2004, Shah 2000
4. Pavitt 1984, Utterback and Abernathy 1978
5. von Hippel 2005a, Berners-Lee 2000
6. Tuomi 2006, Levy 1984
7. Douthwaite 2002
8. von Hippel 2005a
9. Cf. Huppertz 2004
10. Dahlander 2006
11. Huppertz 2004
12. According to the 1890 US Census, farming with 40% of the population, was listed as by far the biggest single occupation. Today, the United Nations Food and Agriculture Organization (FAO) reports that around 1.347 billion people, i.e. given a world population of about 6.4 billion people (UNFPA 2004) that amounts to 21% of the world's total inhabit-

ants, still make up the "economically active population in Agriculture" (FAO 2004).

13. Eurostat 2005. In 1990 farmers constituted less than 2% of the US population (Ikerd 2000).
14. The agricultural sector still employs 40% of workers in developing countries (ILO 2004).
15. Chambers et al. 1989, Richards 1985, Biggs and Clay 1981
16. Jacobson and Smith 2001, Bunch 1989, Rhoades 1989, Chambers et al. 1989, Maurya et al. 1988
17. Biggs and Clay 1981
18. Fowler and Mooney 1990
19. Bellon and Morris 2002
20. Ruthenberg 1985
21. Srinivasan and Thirtle 2000, Richards 1985, Chambers et al. 1989
22. Smith 2005c, Douthwaite 2002
23. Ruthenberg 1985
24. Douthwaite 2002
25. Fernandez-Cornejo 2004
26. Fernandez-Cornejo 2004
27. Koepf and von Plato 2001, Paarlberg and Paarlberg 2000
28. Dixon 1980, Griliches 1964
29. Lu et al. 2004
30. Griliches 1964
31. China accounts for around 30.2% of total global production (Menzel and Tauber 2006).
32. Longping 2004
33. Stadhander 2005, Fowler and Mooney 1990
34. Barnett and Gibson 1999, Maurya 1989
35. Bishay 2003
36. Douthwaite 2002, Srinivasan and Thirtle 2000
37. Fernandez-Cornejo 2004, Chambers et al. 1989
38. Fernandez-Cornejo 2004, Fernandez-Cornejo and Schimmelpfennig 2004
39. Roberts 2004
40. Paarlberg and Paarlberg 2000
41. Cockburn 2004, Kesan 2000
42. BMFT 1994
43. Cockburn 2004
44. Biggs and Clay 1981, Ruthenberg 1985
45. Kesan 2000
46. The International Rice Research Institute (IRRI) in the Philippines for example.
47. Fowler and Mooney (1990), Ruthenberg 1985
48. Von dem Bussche (2004)
49. e.g. From 1800 to 1930 through classical (farmer) breeding methods the yield of winter wheat in Germany grew from 10 dt/ha to 20 dt/ha, i.e. only by 10dt/ha in 130 years. Modern plant breeding methods from the late 1930s/early 1940s increased yields to around 40dt/ha by the late 1960s and to over 70 dt/ha by 1990 (Ordon and Friedt 1998).
50. CGIAR (2004) reports to have developed rice genotypes with 300–400% higher levels of iron, wheat lines with 80–100% more zinc and maize germplasm with 200–275% more provitamin-A than the average.
51. To increase the pest resistance IRRI researchers for instance screened 17,000 varieties and ecotypes of wild rice (Fowler and Mooney 1990).
52. Janaiah 2005
53. Fritschel et al. 1996, cf. Lu et al. 2004

54. PEW 2004
55. Elliesen 2006, PEW 2004
56. Brookes and Barfoot 2006
57. Von dem Bussche 2004
58. Engdahl 2004
59. International Convention for the Protection of New Varieties of Plants of December 2, 1961, as last revised on March 19, 1991.
60. TRIPs requires members to protect plant varieties. In the US the Plant Variety Protection Act (PVPA) was enacted in 1970 to provide protection to novel plant varieties. Genetically engineered produce have been patentable since the US Supreme Court decision *Diamond v. Chakrobarty (1967) 447 US 303, 206 USPQ 193 (1980).*
61. Art. 15, UPOV 1991
62. e.g. 1978 UPOV Convention. Research exemptions allowed protected varieties to be used, without permission, to develop new ones (§ 2544 PVPA).
63. Art. 19 UPOV 1991
64. In UPOV 1991, the right of breeders was extended to the harvested material of the protected variety (Art. 16.2) and the farmer's privilege was made optional (Art. 15.2).
65. Kesan 2000
66. The US research exemption is very limited. In the EU only France and Germany (§ 11 German Patent Law) have a breeding exemption.
67. Salazar et al. 2003
68. UPOV 1991; Art 15.ii and iii
69. Kesan 2000
70. e.g. by requiring all commercialized seeds to be registered, as common in many African countries (Louwaars 2005).
71. EU Regulation 2100/94; § 10a des Sortenschutzgesetz (Plant Variety Protection Law) in Germany.
72. Art. 14 EC Regulation 2100/94; Germany § 10a (2) Sortenschutzgesetz
73. *Asgrow Seed Co. v. Denny Winterboer, 513 US 179 (1995)*.
74. Smith 2005c, Krebs 2004, Coalition Provisional Authority (Iraq) Order Number 81—26 April 2004) § 66.B: "Farmers shall be prohibited from reusing seeds of protected varieties."
75. Krebs 2004, UPOV Convention 1991, Arts. 5–9. In Germany the *Sortenschutzgesetz (SortG)* implements these criteria.
76. Fowler and Mooney 1990, Richards 1985
77. Art. 4 of the Council Directive 2002/53/EC requires that "Member States shall ensure that a variety is accepted only if it is distinct, stable and sufficiently uniform. The variety must be of satisfactory value for cultivation and use."
78. e.g. § 3 of the *Saatgutverkehrsgesetz* (SaatG)
79. BSA 2000
80. Personal interview.
81. While in some cases farmers may obtain access to advanced germplasm, this will often be tied to the acceptance of restrictive license terms (Pardey et al. 2001).
82. PIPRA (2005) stated that "there are only a small number of large multinational firms that control a large proportion of cutting edge agricultural IP."
83. Schimmelpfennig et al. 2004
84. Between 1990–2000 there was an increase of 15% in US biotechnology patents compared to an increase of 5% for patents overall. In Europe, there was a 10.5% increase in biotechnology patents from 1990–1997, compared with a 5% increase overall (OECD 2001).
85. UPOV 2006
86. Kesan 2000

87. PIPRA 2005
88. Ibid.
89. Nash 2000. Ingo Potrykus, one of the two leading scientists who initiated the project, stated that "public sector scientists don't have the capacity to develop a product and bring it to market" (Syngenta 2004).
90. PIPRA 2005, Syngenta 2004
91. Syngenta 2004
92. Louwaars 2005
93. Srinivasan and Thirtle 2000, Fuglie et al. 1996
94. Pollan 1998
95. Roundup Ready Canola seed Technology Use Agreement.
96. *Monsanto v Ralph, 382 F.3d 1374 (2004)*
97. Monsanto allegedly has an annual budget of $10 million dollars and a staff of 75 devoted solely to investigating and prosecuting farmers (CFS 2005).
98. CFS 2005
99. Also see *Monsanto Co. v. McFarling, 302 F.3d 1291 (Fed. Cir. 2002)*; *Monsanto Co., v. Good, 2004 WL 1664013 (D.N.J. 2004)*; Monsanto *v. Scruggs, 2004 WL 1598848 (N.D. Miss 2004)*; *Monsanto Co., v. Swann, 308 F. Supp. 2d 937 (E.D. MO. 2003)*
100. Associated Press 2003
101. CFS 2005
102. Roberts 2004
103. Jacobson and Smith 2001
104. Cf. Bauer 2005
105. Volling 2006, AbL 2006
106. da Silva 2005
107. Bert Scholte in telephone interview; Minot 1986.
108. Fowler and Mooney 1990
109. Such as the Indian or Thai governments (Singh 2005, Chandrasekhar and Ghosh 2003).
110. da Silva 2005 Singh 2005, Minot 1986
111. Chandrasekhar and Ghosh 2003, Fowler and Mooney 1990
112. Macdonald 2004
113. Macdonald and Korb 2006
114. Minot 1986
115. da Silva 2005, Fowler and Mooney 1990
116. Davis 2003, Meek 2000. Pollan 1998
117. Harvey et al. 2002
118. John Deere 2007, ETC Group 2001. Another example is the European Community's use of satellites to verify whether Spanish farmers are complying with the conditions to obtain subsidies In Tasmania a Global Positioning Satellite is being successfully used to assign special identification numbers to agricultural fields to record seed types, fertilizer and spray regimes so that the ultimate purchasers can scrutinize the detailed history of their suppliers. The National Seed Institute of Argentina even proposed to use satellite surveillance to stop illegal seed commerce, increasing the difficulty of commercialising saved seeds (GRAIN 2004, ETC Group 2001).
119. Burk 2004, Srinivasan and Thirtle 2000
120. Burk 2004, Crouch 1998
121. Cf. Crouch 1998 and USDA 2001
122. Le 2005, UNIP Convention on Biological Diversity 2003. The US however is not a signatory to the Convention.
123. e.g. The International Seed Foundation, and PHARMA.

124. Canada, New Zealand and Australia, as well as the US Department of Agriculture (Le 2005, cf. USDA 2001).
125. Jorash 2006, Burk 2004
126. Alone between 1999 to 2007 the number of farms declined from 434,130 to 353,300 (BMELV 2007a and 2007b).
127. BMELV 2007c
128. e.g. telephone interviews with Mr. Werner Bosse and Dr. Wolfgang Sohn, Farmer Association of Lower Saxony; Mr. Wilhelm Grimm, President Farmer Association of Hamburg; Dr. Carola Bühler, Agricultural Support Office, Hamburg Chamber for Agriculture); Mr. Leonard Keller, President of the Bavarian Farmer Association for the region Schwaben; Mrs. Newe, Plantproduction Expert for the Farmer Association of Schleswig-Holstein; Mrs. Belinda Giesen-Druse, responsible for crop seeds at the Association of German Plantbreeders (BDP): and Mrs. Ulrike Jungmann, Communications Officer of the BDP.
129. Around 50% of seeds are still saved each year in Germany (Elliesen 2006).
130. Lambke et al. 2003
131. Cf. Monitor 2005
132. Schievelbein 2000
133. Werner Bosse
134. Mr. Keller and Mrs. Belinda Giesen-Druse emphasized this.
135. Lammerts von Bueren 2002
136. Rohner-Thielen 2005
137. Most European governments intend organic agriculture to account for up to 10% of total agriculture by 2010 (Lammerts von Bueren 2002).
138. Willer and Yussefi 2006
139. Lammerts von Bueren 2002
140. Koepf and von Plato 2001
141. Demeter 2005. South German farmer Peter Jacoby for example developed 2 improved organic wheat varieties by crossing old landraces (Koepf and von Plato 2001).
142. Lammerts von Bueren 2002
143. Koepf and von Plato 2001
144. www.kartoffelvielfalt.de
145. Personal interview with Karsten Ellenberg, October 9, 2006.
146. Irle 2005
147. Email correspondence with Karsten Ellenberg, October 23, 2006.
148. Lammerts van Bueren 2002
149. Email correspondence with Karsten Ellenberg, October 23, 2006.
150. Smith 2005c, Douthwaite 2002
151. Tuxill 1999
152. Ibid.
153. Virmani 1998
154. Maurya et al. 1988, Biggs and Clay 1981
155. e.g. conditions in Great Britain are similar to those in Germany, the Netherlands or Northern France. Farming conditions are also quite homogenous in the "Grain Belt" of the Midwest where the majority of US grain is grown.
156. Bellon and Morris 2002, Maurya et al. 1988
157. Pardey et al. 2001
158. Ibid.
159. Ibid.
160. In South and West Asia and Sub-Saharan Africa illiteracy rates are still around 40% (UNESCO 2005).
161. Vernooy 2003
162. Douthwaite 2002

163. von Hippel 1976
164. Berger 1975
165. Boyden 1976
166. Cf. Rogers 2003
167. Rogers 2003
168. Rice and Rogers 1980
169. Werker and Athreye 2004, Nelson 1993
170. Fritsch and Slavtchev 2005
171. Romer 1986
172. Audretsch and Feldman 1996, Krugman 1991
173. Anselin et al. 2000
174. BMBF 2005
175. Nuvolari 2004, Macleod 1988, Allen 1983
176. Petrash and Bukowitz 1997
177. Lerner and Tirole 2002, Rothaermel and Sugiyama 2001
178. Kline and Pinch 1994
179. Franz 1999
180. The Economist 2006b, The Economist 2006c.
181. Chen and Liu 2005
182. Cohen et al. 2000, Mansfield 1986
183. Rogers 2003, Varian 2002
184. Virmani et al. 2002, Douthwaite 2002
185. Vonortas and Kim 2004, Reitzig 2004, Cohen et al. 2002
186. Cf. Levin et al. 1987
187. Bessen and Meurer 2005
188. von Hippel and Katz 2002
189. Leonard Keller
190. Graham and Mowery 2004, Rogers 2003
191. Lotz 1992
192. BMBF 2005
193. Shapiro 2000
194. Galli 2005
195. Benkler 2002
196. Cohen and Levinthal 1990
197. Nelson 1996
198. Aumiller 1973
199. BMBF 2005
200. Cf. Biemans 1991
201. von Hippel 2005a
202. Overall in these 4 studies 74 users developed one major innovation, 4 developed more than 1 and only 2 of these more than 2.
203. Cohen and Levinthal 1990
204. Berners-Lee 2000, Shapiro 2000

## PART D

1. Spilgies and Möhrle 2006
2. The *New Oxford Dictionary of English* defines a license as "a permit from an authority to own or use something, to do a particular thing, or to carry on a trade" (Pearsall 1998).
3. Duffus 2003
4. Varian 2002, Samuelson 1999
5. Garfinkel 1995

6. This deal subsequently broke down as the relations between RsA and Cylink deteriorated in 1994 (Anthes 1997).
7. Cf. Granstrand 1999
8. Vonortas and Kim 2004, Yang and Maskus 2001, Gans et al. 2000, Somaya and Teece 2000
9. Carrier 2003
10. Graham and Mowery 2004, Shapiro 2001, Hall and Ziedonis 2001, Granstrand 1999, Grindley and Teece 1997
11. Cf. Lessig 2004
12. Harhoff et al. 2003
13. e.g. Such as IBM (hardware) and Red Hat (service) in embracing the open-source movement (cf. von Hippel 2005b, Allen 1983). Lerner and Tirole (2002) explained that through free revealing OSS programmers can increase their value on the job market.
14. e.g. checking for software bugs or allowing third parties to improve a product that is tailored to one's internal processes.
15. For example, see Poynder (2003b) who argues that the most valuable content frequently ends up in the ownership of large media and information companies.
16. Pharmalicensing.com for example provides information about licensing transactions in the pharma industry. Each report costs many thousands of dollars (Pharmalicensing 2006).
17. Directive 96/9
18. Dutfield and Suthersanen 2004
19. www.creativecommons.org
20. Lim 2004
21. Nuvolari (2004) explains that collective invention involves firms publicly disclosing their incremental inventions and thereby forming a common pool of technological knowledge to further improve the technology in question.
22. Cf. Schiffer 2003, Macleod 1988
23. Harhoff et al. 2003
24. Henkel and von Hippel 2005, Henkel 2004
25. US patent law states that an innovator has 1 year from the date an enabling disclosure is made to file a patent application. In most other countries an innovator may not apply for a patent after an enabling disclosure has been made.
26. Lee et al. 2003, Lerner and Tirole 2002
27. Hope 2004
28. Razgaitis 2003, Mishina 1989
29. Brodkin 2008
30. LaMonica 2008
31. www.wbcsd.org/web/epc/ (As of April 17, 2008)
32. Tuomi 2006, Dahlander 2004
33. Williams 2005
34. Dejong 2005, IBM Annual Report 2004
35. Cf. Vonortas and Kim 2004
36. Evans 2003
37. e.g. von Hippel (1988; pp. 48–52): "In sum, empirical data seem to suggest that the patent grant has little value to innovators in most fields."
38. Bessy and Brousseau 2000, Levin et al. 1987
39. Gabrielle Campbell made this point in an email correspondence with one of the authors.
40. *Madey vs. Duke University (Federal Circuit 01–1567) October 3, 2002*
41. *Embrex, Inc. v. Service Engineering Corp 216 F.3d 1343 Fed. Cir. 2000*
42. Heller and Eisenberg 1998
43. NIH 1998

44. Chesbrough 2003
45. Walsh et al. 2003, Heller and Eisenberg 1998
46. Price 1999
47. National Science Board 2004, Bok 2003
48. AUTM 2004
49. Slaughter and Leslie 1997
50. Bok 2003
51. Rai 1999, NIH 1998
52. Washburn 2005; Jaffe and Lerner 2004
53. Long 2001
54. National Science Board 2004
55. In the US biotech industry research spending amounts to approximately 80% of the sector's total revenues, compared to around 5% in electronics (Suarez-Villa 2004).
56. Suarez-Villa 2004, Cavalla 1997
57. Biemans 1991, Shaw 1985
58. Lotz 1992
59. Prahalad and Ramaswamy 2004
60. IBM 2006
61. Keegan 2001a, 2001b
62. Trott 1998
63. Hammann and Mittag 1986
64. Böll 2001
65. Cf. Bills 2004
66. Cf. Telesio 1977
67. van Dijk 2000
68. Anand and Khanna 2000
69. Cf. Long 2001, Macleod 1988
70. Vonortas and Kim 2004, Arora et al. 2002, Anand and Khanna 2000
71. Grindley and Teece 1997
72. Contractor 1999
73. Martinez et al. 2004
74. Mori 2005, Razgaitis 2004, Chesbrough 2003, Arora et al. 2002, LER 2001
75. Contractor and Lorange 2002
76. Arora et al. 2002
77. Gans et al. (2000) found empirical support for the assertion that stronger IPRs are associated with higher levels of co-operation, often in the form of licenses, between incumbents and start-up innovators.
78. Suarez-Villa 2003, Chesbrough 2003, Contractor and Lorange 2002, Arora et al. 2002
79. Martinez et al. 2004
80. Foxall and Johnston 1987
81. e.g. Motohashi 2006, Kollmer and Dowling 2004, Hall and Ziedonis 2001, Grindley and Teece 1997, Hamman and Mittag 1986, Gallini 1984, Greipl and Täger 1982, Wilson 1977, Lovell 1969
82. McDermott and O'Connor 2002, Comerford 1984, Greipl and Täger 1982
83. Mordhorst 1994, Greipl and Täger 1982
84. Howells 1999
85. Cf. Engardio and Einhorn 2005
86. Pastor and Sandonis 2002, Milston et al. 1996, Mowery 1992
87. Thirteen percent contained property rights that had to be granted back, 47.8% contained use rights and 4.4% simple information rights.
88. Degnan and Horton 1997. Another LES Study of 229 US/Canadian LES members found that 41% of out-licensing agreements and 45% of in licensing agreements contained grant-back provisions (Razgaitis 2004).

89. Bessy et al. (2002) found 46.9%. In Degnan and Horton 's 1997 study of 428 LES members 80% of licensing agreements contained a provision for improvements.
90. Kasarda and Rondinelli 1998
91. This case study is mainly based on interviews with the manufacturer.
92. Doyle 2000
93. Guinnessy 1997
94. Bühner 1997
95. Doyle 2000
96. Harrison 2004, www.xerox.com
97. Jaffe and Lerner 2004
98. Tibballs 1994
99. www.xerox.com, Tibballs 1994
100. Tibballs 1994
101. Jaffe and Lerner 2004
102. Cameron and de Vries 2006, Douthwaite 2002, Jorgensen 1997
103. Cameron and de Vries 2006
104. Wüstenhagen 2003
105. Energy Economist 2005, Jorgensen 1997
106. Lotz 1992
107. Wible 2003
108. Mollick 2004
109. Keegan 2001a, 2001b
110. Larry Keeley (2005), Professor at the Institute of Design at Illinois Institute of Technology, stated that: "I have been toiling intensively to separate innovation from design for over a decade."
111. Razgaitis 2004, Duysters et al. 2000
112. Tietz et al. 2006, Urban and von Hippel 1988
113. Tietz et al. 2006, von Hippel et al. 2000
114. Tietz et al. 2006
115. Reichwald and Piller 2006
116. Arora and Fosfuri 2003, Duysters et al. 2000
117. Boller 2007
118. Bessy and Brousseau 2000
119. Matsunaga 1977
120. Gawer and Cusumano 2002
121. Cf. Herstatt and Lettl 2004
122. Nagel 1993, Mishina 1989
123. Arora 1996, Macho-Stadler et al. 1996
124. Spilgies and Möhrle 2006, Goedkoop et al. 1999
125. von Hippel 2005a, Lüthje et al. 2005
126. Chesbrough 2003
127. BMBF 2005
128. Prahalad and Ramaswamy 2004
129. Samuelson and Scotchmer 2002
130. von Hippel and Katz 2002
131. Vernooy 2003, Maurya 1989, Richards 1985
132. Richards 1985
133. Douthwaite 2002
134. Chambers 1983
135. Chambers et al. 1989
136. Sperling et al. 1996
137. Richards 1985
138. Bellon and Morris 2002

139. Vernooy 2003, Maurya 1989
140. Witcombe 2005
141. Douthwaite 2002
142. Witcombe 2005
143. Frossard 1995
144. Douthwaite 2002
145. Frossard 1995
146. Virmani et al. 2002
147. Witcombe 2005
148. Vernooy 2003
149. Witcombe 2005
150. Bellon and Morris 2002
151. Salazar et al. 2003
152. Lammerts von Bueren 2002
153. www.meijer-potato.com
154. http://www.theoscarproject.org
155. Winston 2005, Lehman 2006
156. See the case of chainsaw manufacturer Stihl (Maier 2006).

# Bibliography


AAIA (2006) "Car company vehicle service information & tools: An evaluation of their availability to independent repair shops" Automotive Aftermarket Industry Association.

Abernathy, W., and Utterback, J. (1978) "Patterns of industrial innovation" *Technology Review* 80(7): 40–47.

AbL (2006) *Monsanto's Genmais—Anbauvertrag in Deutschland.* Hamm/ Lüneburg, Arbeitsgemeinschaft bäuerliche Landwirtschaft.

ActionAid (2005) *Power hungry: Six reasons to regulate global food corporations*, ActionAid International.

Adobe (2007) "End-user license agreement for Adobe Reader" www.adobe.com/ products/acrobat/acrreula.html—last accessed May 14, 2007.

Agarwala-Rogers, R. (1978) "Re-invention: The reshaping of innovations in adoption" *Organizational Buying Behavior*: 138–145.

AIPLA (2003) Report of the Economic Survey 2003. Washington, D.C., American Intellectual Property Law Association.

Allen, R. (1983) "Collective invention" *Journal of Economic Behavior and Organization* 4(1): 1–24.

Allison, J. and Lemley, M. (200) "Who's Patenting What? An Emperical Exploration of Patent Peosecution", *Baolt Working Papers in Public Law* (98)

Allison, J. and Lemley, M. (2001) "The growing complexity of the United States patent system" *UC Berkeley Public*, Law Research Paper No. 66.

Anand, B. and Khanna, T: (2000) "The structure of licensing contracts" *The Journal of Industrial Economics* 48(1): 103–135.

Anderson, R. (2003a) "Trusted computing" Frequently asked questions www. cl.cam.ac.uk/users/rja14/tcpa-faq.html—last accessed May 23, 2007.

Anderson, R. (2003b) "'Trusted computing' and competition policy—Issues for computing professionals" *UPGRADE* IV(3): 35–41.

Anneken, T. (2005) *Vergleich verschiedener Puder- und Lack Systeme zur optischen Kavitätenvermessung bei CDENT 3D* Medizinische Fakultät, Universität Hamburg, Hamburg.

Anselin, L., Varga, A. and Acs, Z. (2000) "Geographical spillovers and university research: a spatial econometric perspective" *Growth and change* 31(4): 501–515.

Anthes, G. (1997) "Encryption dispute settled" *Computerworld* 31(2): 32.

Antitrust Guidelines for the Licensing of Intellectual Property, issued by the US Department of Justice and the Federal Trade Commission, April 6, 1995.

Arai, H. (1997) *Towards the era of intellectual creation: IPR protection in the 21st century*, Japanese Patent Office.

Arbaugh, W., Farber, D., and Smith, J. (1997) *A secure and reliable bootstrap architecture.* IEEE Symposium on Security and Privacy, IEEE.

Archibugi, D., Cesaratto, S., and Sirelli, S. (1986) *Firm Size and Sources of Innovation: An Analysis of the Italian Survey on Innovation Diffusion*. OECD Workshop on Innovations Statistics, December 8–9, 1986, Paris, Organisation for Economic Co-operation and Development.

Ardisson, J.-M., and Bidault, F. (1986) "Technology transfer strategies as a means to build international networks" *Industrial Marketing & Purchasing* 1(1): 59–72.

Arora, A. (1996) "Contracting for tacit knowledge: The provision of technical service in technology licensing contracts" *Journal of Development Economics* 50: 233–256.

Arora, A. (1997) "Patents, licensing and market structure in the chemical industry" *Research Policy* 26(4): 391–403.

Arora, A., and Fosfuri, A. (2003) "Licensing the market for technology" *Journal of Economic Behavior and Organization* 52: 277–295.

Arora, A., Ceccagnoli, M., and Cohen, W. (2003) *R&D and the patent premium*, Working Paper 9431. Cambridge, MA, National Bureau of Economic Research.

Arora, A., Fosfuri, A. and Gambardella, A. (2002) *Markets for technology: The economics of innovation and corporate strategy*. London, MIT Press.

Arundel, A. (2001) "The relative effectiveness of patents and secrecy for appropriation" *Research Policy* 30(4): 611–624.

Ashley, S. (1993) "High-tech rackets hold court" *Mechanical Engineering* 115(8): 50–55.

Ashley, S. (1995) "A new racket shakes up tennis" *Mechanical Engineering* 117(8): 80–81.

Associated Press (2003) "Farmer sent to prison over cotton seed" *Associated Press* (May 10, 2003).

Åstebro, T. (1998) "Basic statistics on the success rate and profits for independent inventors" *Entrepreneurship Theory and Practice* 23(2): 41–48.

Atkinson, R., Beachy, R., Conway, G., Cordova, F. et al. (2003) "Public sector collaboration for agricultural IP management" *Science* 301(5630): 174–175.

Audretsch, D., Feldman, M. (1996) "R&D spillovers and the geography of innovation and production" *The American Economic Review* 86(3): 630–640.

Aumiller, J. (1973) *Ingenieure verändern die Medizin*. Freiburg, Herder Druck.

AUTM (2004) *AUTM Licensing Survey: FY 2003*, The Association of University Technology Managers.

AUTM (2007) *AUTM Licensing Survey: FY 2005*, The Association of University Technology Managers.

Babcock, C., and Foley, J. (2004) "The cost of ideas" *Information Week* (1009): 20–22.

Bahney, A. (2006). "High tech, under the skin" *New York Times* New York, (February 2, 2006):G.1.

Bailey, R. (2007) "Profiting from the infinitesimal" *Chief Executive*. New York: July/August 2007, 227: 34

Baldwin, C., and Clark, K. (2004) "Managing in an age of modularity" *Managing strategic innovation and change: A collection of readings*. Tushman, M. and Anderson, P. New York, Oxford University Press: 151–160.

Baldwin, C., Hienerth, C., von Hippel, E. (2006) "How user innovations become commercial products: A theoretical investigation and case study" *Research Policy* 32(1): 157–178.

Band, J. (2005) "Caught in a blizzard" *Journal of Internet Law* 9(5): 3–5.

Barnett, B., and Gibson, B. (1999) "Economic challenges of transgenic crops: The case of BT cotton" *Journal of Economic Issues* 33(3): 647–659.

Barr, M. (2003) "Democracy in the dark: Public access restrictions from Westlaw and LexisNexis" *Searcher* 11(1): 66–72.

Barro, R., and Sala—I—Martin, X. (1995) *Technology diffusion, convergence, and growth*, Working paper 5151, Cambridge, MA, National Bureau of Economic Research.

Bartlett Foote, S. (1991) "The Impact of Public Policy on Medical Device Innovation: A Case of Polyintervention" *The Changing Economics of Medical Technology*. A. Gelijns, and Halm, E. Washington, D.C, National Academy Press: 69–88.

Bauer, A. (2005) "Eine ökonomische Katastrophe" *Umweltnachrichten* 102 (December 2005).

Baumol, W. (2004) "Four sources of innovation and stimulation of growth in the Dutch economy" *Fostering productivity: Patterns, determinants and policy implications*. G. Gelauff, Klomp, L., Raes, S., and Roelandt, T. Amsterdam: 185–198.

Bayer Annual Report (2003) Bayer Annual Report 2003 www.bayer.com

BayNews (2006) "adidas und Bayer MaterialScience im Doppelpass bei der Entwicklung des FIFA WM—Balls: Nie war er so rund wie heute" *BayNews* (March 15, 2006)

BBC (2006). "Half of UK 'infringing copyright" *BBC* (May 12, 2006)

Bechtold, S. (2004) "Trusted computing initiatives—Protecting virtual Troy or creating a Troyan Horse?" *Trusted computing*. C. Koenig, Neumann, A., and Katzschmann, T. Heidelberg, Verlag Recht und Wirtschaft.

Behrman, J. (1958) "Advantages and disadvantages of foreign licensing" *The Patent, Trademark and Copyright Journal of Research and Education* 2(1).

Bellon, M., and Morris, M. (2002) *Linking global and local approaches to agricultural technology development: The role of participatory plant breeding research in the CGIAR*. Mexico D.C., CIMMYT.

Bellovin, S., Blaze, M., Boneh, D., Del Torto, D., Goldberg, B., Schneier, B., Stevenson, F., and Wagner, D. (2001) "Brief of Amici Curiae in support of appellants and reversal of the judgment in *Universal v. Reimerdes*" (January 26, 2001). *United States Court of Appeals for the Second Circuit*.

Benkler, Y. (2002) "Intellectual property and the organization of information production" *International Review of Law and Economics* 22(1): 81–107.

Berger, A. (1975) "Factors influencing the locus of innovation activity leading to scientific instrument and plastics innovations. *Sloan School of Management*. Cambridge, MA, MIT.

Bergstein, B. (2003) "Encryption revolution: The tantalizing promise of 'unbreakable' codes" *Detroit News*. Detroit: web.

Berkowitz, L. (2003) "'Research exemption' to patent infringement narrowing" *Research Technology Management* 46(3): 4.

Berners-Lee, T. (2000) *Weaving the web: The past, present and future of the world wide web by its inventor*. London, Texere.

Bernstein, A. (2000) "Too much corporate power?" *Business Week* (September 11, 2000).

Bessen, J., and Hunt, R. (2004) "The software patent experiment" *Business Review* 3: 22–32.

Bessen, J., and Meurer, M. (2005) The patent litigation explosion. Boston, Boston Univ. School of Law Working Paper No. 05–18.

Bessy, C., and Brousseau, E. (1998) "Technology licensing contracts: Features and diversity" *International Review of Law and Economics* 18 (December 1998): 451–489.

Bessy, C., and Brousseau, E. (2000) "Technology licensing practices in France" *LES Nouvelles* (June 2000): 66–75.

Bessy, C., Brousseau, E., and Saussier, S. (2002) "The diversity of technology licensing agreements" Paris, ATOM (Université de Paris I): 1–30.

Biemans, W. (1991) "User and third-party involvement in developing medical equipment innovations" *Technovation* 11(3): 163–182.

Biggs, S., and Clay, E. (1981) "Sources of innovation in agricultural technology" *World Development* 9(4): 321–336.

Bills, K. (2004) "A guide to licensing biotechnology" *LES Nouvelles* (June 2004): 86–94.

Bishay, F. (2003) *Towards sustainable agricultural development in Iraq* Rome, Food and Agriculture Organization of the United Nations.

Blair, M., and Wallman, S. (2001) *Unseen wealth: Report of the Brookings Task Force on Intangibles.* Washington, D.C., Brookings Institution Press.

Blind, K., Edler, J., Nack, R., and Strauss, J. (2003) *Mikro- und makroökonomische Implikationen der Patentierbarkeit von Softwareinnovationen: Geistige Eigentumsrechte in der Informationstechnik im Spannungsfeld von Wettbewerb und Innovation.* Karlsruhe, German Ministry for Economics and Technology.

Blume, S. (1995) "Cochlear implantation: Establishing clinical feasibility, 1957–1982" *Sources of Medical Technology: Universities and Industry.* A. Gelijns, and Rosenberg, N., Institute of Medicine: 97–124.

BMBF (2005) *Studie zur Situation der Medizintechnik in Deutschland im internationalen Vergleich.* Berlin, Bundesministerium für Bildung und Forschung.

BMELV (2007a) "Landwirtschaftliche Betriebe und Ihre Flächen" Bundesministerium für Ernährung, Landwirtschaft und Verbraucher, Statistik, http://www.bmelv-statistik.de/tabellen/f1060.1.xls—last accessed January 16, 2008.

BMELV (2007b) *Agrarpolitischer Bericht der Bundesregierung 2007* Berlin, Bundesministerium für Ernährung, Landwirtschaft und Verbraucher.

BMELV (2007c) "Wertschöpfung der Landwirtschaft" Bundesministerium für Ernährung, Landwirtschaft und Verbraucher, Statistik, http://www.bmelv-statistik.de/tabellen/f1540.1.xls—last accessed March 1, 2007.

BMFT (1994) *Biotechnologie in der Pflanzenzüchtung: Einführung in die Thematik, Aspekte der BMFT-Forschungsförderung, Berichte aus Forschungsprojekten.* Bonn, Bundesministerium für Forschung und Technologie.

Bock, L. (2004) "Following Mr. Robinson's Advice: The Story of Nanosys" *Nanotechnology Law & Business Journal* 1 (1, Art. 10): 1–7.

Bodoni, S. (2005) "How an IP policy can help your company grow" *Managing Intellectual Property* (147): 30–35.

Boeckmann, R.-D., and Frankenberger, H. (2002) *MPG & Co: Eine Vorschriftensammlung zum Medizinprodukterecht mit Fachwörterbuch.* Köln, TÜV Verlag.

Bok, D. (2003) *Universities in the market place.* Princeton, Princeton University Press.

Bolan, C. (2005) "How to catch a counterfeiter" *Global Cosmetic Industry* 173(6): 38–40.

Böll, K., Ed. (2001) *Handbuch licensing.* Frankfurt am Main, Deutscher Fachverlag.

Boller, W. (2007) "Tognum: Stechen Finnen Börse aus?" *Südkurier* (June 14, 2007).

Borenstein, S. (2000) "Exercising market power in proprietary aftermarkets" *Journal of Economics and Management Strategy* 9(2): 157–188.

Borland, J. (2003) "Court blocks security conference talk" *CNET News* (April 14, 2003)

Borland, J. (2004a) "New CD copy-lock technology nears market" *CNET News* (December 16, 2004)

Borland, J. (2004b) "Judge rules DVD-copying programme is illegal" *CNET News.com* (February 23, 2004).

Boskin, M., and Lau, L. (2000) "Generalized solow-neutral technical progress and postwar economic crowth" National Bureau of Economic Research, NBER Working Paper 8023.

Bovens, A. (2005) "Closed architectures for content distribution" *Japan Media Review* (February 12, 2005).

Bowman, L. (2002) "Font companies crack down on student" *News.com.* (May 3, 2002)

Boyden, J. (1976) *A study of the innovative process in the Plastics Additives Industry* Sloan School of Management. Cambridge, MA, MIT.

Braham, J. (1995) "Boom, boom, boom! And Bertha's even bigger!" *Machine Design* 67(16): 36–40.

Bramwell, T. (2006) "Microsoft and hackers dismiss danger of Xbox360 security" *Gamesindustry.biz* (March 24, 2006).

Branscomb, L., and Kodama, F. (1993) *Japanese innovation strategy.* Cambridge, MA, Centre for Science and International Affairs.

Brinton, J. (1993) "Technology transfer and commercialization" *Understanding biotechnology law—Protection, licensing and intellectual property policies.* G. Peterson. New York, Dekker: 431–455.

Brodkin, J. (2008) "IBM retains slim lead among patent—winners IBM followed closely by Samsung, Canon, Panasonic and Intel" *Computer World*, January 14, 2008.

Brookes, G., and Barfoot, P. (2006) "GM crops: the first 10 years—Global socio—Economic and environmental impacts" *ISAAA Brief*, No. 36.

Brown, K. (2004) "Outsourcing and the devaluation of intellectual property" *Darwin* (April 26, 2004).

Bruns, B. (2001) "Open sourcing nanotechnology research and development: Issues and opportunities" *Nanotechnology* 12: 198–210.

BSA (Business Software Alliance) and IDC (International Data Corporation) (2006) *Third annual BSA and IDC global software piracy study* May 2006.

Bühner, J. (1997) "Das CEREC-system: Der Einsatz von CEREC in zahnärztlichen Praxis" *Medizinische Fakultät.* Tübingen, Eberhard-Karls-Universität.

Bulman, R., and Lehman, J. (2004) "Attacking the hacking from inside" *Journal of Internet Law* 8(6): 16–20.

Bunch, R. (1989) "Encouraging farmers' experiments" *Farmer first—Farmer innovation and agricultural research.* R. Chambers, Pacey, A., and Thrupp, L.A. London, Intermediate Technology Publications: 55–61.

Burk, D. (2004) "DNA rules: Legal and conceptual implications of biological "lock-out" systems" *California Law Review* 92: 1553–1587.

Butcher, M. (2004) "Music to the ears" *The Guardian* (Thursday June 10, 2004).

Cameron, A., and de Vries, E. (2006) "Top of the list: A quick look at how the major turbine companies fared in 2005 and what's in store for 2006" *Renewable Energy World* 9(1): 28–41

Carrier, M. (2003) "Resolving the patent-antitrust paradox through tripartite innovation" *Vanderbilt Law Review* 56(4): 1056.

Cavalla, D. (1997) *Modern strategy for pre-clinical pharmaceutical R&D: Towards the virtual research company.* Chichester, John Wiley and Sons.

Caves, R., Crookel, H., and Killing, J. (1983) "The imperfect market for technology licenses" *Oxford Bulletin of Economics and Statistics* (45): 249–267.

CEREC Forum (2004) Powder for your CEREC, some impressions, http://www.cerec.net/forum/index.php?s=1191314e7c661c749e9f1f61359e1637&showtopic=489—last accessed January 17, 2007.

CFS (2005) *Monsanto vs. US Farmers.* Washington D.C., Centre for Food Safety.

CGIAR (2004) *CGIAR annual report 2004—Innovations in agricultural research*, Consultative Group on International Agricultural Research (CGIAR).

Chambers, R. (1983) *Rural development: Putting the last first.* London, Longman.

Chambers, R., Pacey, A., and Thrupp, L., Ed. (1989) *Farmer first—Farmer innovation and agricultural research.* London, Intermediate Technology Publications.

Chandrasekhar, C., and Ghosh, J. (2003) "Is corporate farming really the solution for Indian agriculture?" *The Hindu BusinessLine* (December 16, 2003).

Chandy, R., and Tellis, G. (1998) "Organizing for radical product innovation: The overlooked role of willingness to cannibalize" *Journal of Marketing Research*, 35(4): 474–487

Chaponniere, J., and Lautier, M. (1995) "Breaking into the Korean Market—Invest or license?" *Long Range Planning* 28(1): 104–111.

Chartrand, S. (2004) "Does the patent system need an overhaul?" *The New York Times.* New York (September 27, 2004).

Chen, K.–M., and Liu, R–J. (2005) "Interface strategies in modular product innovation" *Technovation* 25: 771–782.

Chesbrough, H. (2003) *Open innovation—The new imperative for creating and profiting from technology.* Boston, Harvard Business School Press.

Chesbrough, H., and Kardon Crowther, A. (2006) "Beyond high tech: Early adopters of open innovation" *Research Policy* 36(3): 229–236.

Chevigny, P. (1965) "The validity of grant-back agreements under the antitrust laws" *Fordham Law Review* 34: 569.

Choi, C. (2004) "Nano World: Nanotech patent wars may come" UPI (United Press International) (October 1, 2004).

Choi, J. (2002) "A dynamic analysis of licensing: The "boomerang" effect and grant-back clauses" *International Economic Review* 43(4): 803–826.

Christensen, C. (1997) *The innovator's dilemma: When new technologies cause great firms to fail.* Boston, MA, Harvard Business School Press.

Cockburn, I. (2004) "The Changing Structure of the Pharmaceutical Industry", *Health Affairs*, 23(1): 10-22

Cockburn, I., and Henderson, R. (2003) *Survey results from the 2003 Intellectual Property Owners Association survey on strategic management of intellectual property*, IPO (Intellectual Property Owners Association).

Cohen, A. (2006) "Microsoft ends appeal of penalty levied by EU" *The Wall Street Journal* (May 2, 2006): 6.

Cohen, J., and Lemley, M. (2001) "Patent scope and innovation in the software industry" *California Law Review* 89: 1.

Cohen, W., and Levinthal, D. (1990) "Absorptive capacity: A new perspective on learning and innovation" *Administrative Science Quarterly* 35: 128–152.

Cohen, W., Goto, A., Nagata, A., Nelson, R., and Walsh, J. (2002) "R&D spillovers, patents and the incentives to innovate in Japan and the United States" *Research Policy* 31(8–9): 1349–1367.

Cohen, W., Nelson, R. and Walsh, J. (2000) "Protecting their intellectual assets: Appropriability conditions and why US manufacturing firms patent (or not)" Working Paper no. 7552. Cambridge, MA, National Bureau of Economic Research.

Coleman, J., Katz, E. and Menzel, H. (1964) "The diffusion of an innovation among physicians" *Social Processes and Structures*: 100–117.

Comerford, K. (1984) *Licensing and finding new products* Dublin, Institute for Industrial Research & Standards.

Computer and Internet Lawyer (2004) "Kodak and Sun Microsystems settle patent infringement lawsuit" *Computer and Internet Lawyer* 21(12): 20.

Computer and Internet Lawyer (2005) "Federal circuit overturns verdict in Eolas vs. Microsoft patent infringement case" *Computer and Internet Lawyer* 22(5): 21.

Computerbase (2005) "Mod–chips für Xbox 360 in Arbeit" http://www.computerbase.de/news/consumer_electronics/konsolen/microsoft/2005/dezember/mod-chips_xbox_360_arbeit—last accessed February 20, 2007.

Conner, M. (2006) "Friend or foe: Battery-authentication ICs separate the good guys from the bad" *EDN* (February 2, 2006).

Contractor, F. (1985) *Licensing in international strategy.* Westport, Connecticut, Quarum Books.

Contractor, F. (1999) "Licensing compared with foreign equity investment and trading for US-based multinational firms" *Journal of the Licensing Executives Society* (June 1999): 1–10.

Contractor, F., and Lorange, P., Ed. (2002) *Cooperative strategies and alliances.* International Business and Management Series, Pergamon.

Cooper, R., and Kleinschmidt, E. (1990) *New products: The key factors in success.* Chicago, American Marketing Association.

Coopers & Lybrand (1997) *Maximizing the value of intellectual property*, Coopers & Lybrand.

Crouch, M. (1998) *How the terminator terminates: An explanation for the non-scientist of a remarkable patent for killing second generation seeds of crop plants*. Bloomington, Indiana, The Edmonds Institute.

Cullet, P. (2004) *Plant variety protection* IELRC Briefing Paper 2003–4, International Environmental Law Research Centre.

Curran, W. (1984) "The unwanted suitor: Law and the use of health care technology" *Machine at the bedside*. S. Reiser, and Anbar, M. Cambridge, Cambridge University Press.

Czarnota, B., and Hart, R. (1991) *Legal protection of computer programs in Europe: A guide to the EC directive*, Butterworths Law.

da Silva, C. (2005) *The growing role of contract farming in Agri–Food systems development: Drivers, theory and practice*. Rome, FAO.

Dahlander, L. (2004) *Appropriating returns from open innovation processes: A multiple case study of small firms in open source software*, Working Paper Chalmers University of Technology, Gothenburg.

Dahlander, L. (2006) "Managing beyond firm boundaries: Leveraging user innovation networks" *Department of Technology Management and Economics*. Gothenburg, Chalmers University of Technology.

Dahlander, L., and Magnusson, M. (2005) "Relationships between open source software companies and communities: Observations from Nordic firms" *Research Policy* 34(4): 481–493.

Dahlander, L., and McKelvey, M. (2005) "Who's not developing open source software? Non-users, users, and developers" *Economics of Innovation and New Technology* 14(7): 617–635.

Dahlin, K., Taylor, M., and Fichman, M. (2004) "Today's Edisons or weekend hobbyists: Technical merit and success of inventions by independent inventors" *Research Policy* 33: 1167–1183.

D'Aveni, R. (1994) *Hypercompetition: Managing the dynamics of strategic manoeuvring*. New York, NY, Free Press.

Davidson, P. (2004) "Patents out of control? Growing lawsuits shake up Internet industry" *USA Today* (January 13, 2004): B.01.

Davis, K. (2003) *Genetic engineering and cloning of domestic fowl*. FDA Fall Public Meeting: Issues Relating to Animals Raised for Food, United Poultry Concerns.

De Grauwe, P., and Camerman, F. (2003) "Are multinationals really bigger than nations?" *World Economics* 4(2): 23–37.

Degnan, S. A., and Horton, C. (1997) "A survey of licensed royalties" *LES Nouvelles* (June 1997): 91–96.

Dejong, J. (2005) "Not so patently clear" *Software Development Times* 123(April 1, 2005): 1–2.

Dell (2007) Dell's software license agreement http://www1.jp.dell.com/content/topics/global.aspx/policy/en/policy?c=jp&l=jp&s=corp&~section=015—last accessed March 12, 2007.

Demeter (2005) "Demeter gratuliert seinem Förderpreisträger" Demeter.

DeMonaco, H., Ali, A., and von Hippel, E. (2005) *The major role of clinicians in the discovery of off-label drug therapies*, MIT Sloan School of Management.

Devil 360 (2006) "Devil 360: Xbox 360 Security Info", www.xboxscene.com/xbox1data/sep/EEuVVAEppFyjtTTvtlphp - last accessed December 8, 2008

Diamond, J. (1998) *Guns, germs and steel*. London, Vintage.

Dixon, R. (1980) "Hybrid corn revisited" *Econometrica* 48(6): 1451–1461.

Dornan, A. (2006) "'Trusted network connect' puts hardware security agent in every PC" *InformationWeek* (February 14, 2006).

Douthwaite, B. (2002) *Enabling innovation: A practical guide to understanding and fostering technological change*. London, Zed Books.

Doyle, A. (2000) "Digital dentistry" *Computer Graphics World* 23(10): 50–53.

Duffus, A. (2003) "How Europe is striving for copyright harmony" *Managing Intellectual Property* (December 2003).

Dusollier, S. (2003) "Fair use by design in the European copyright directive of 2001" *Communications of the ACM archive* 46(4): 51–55.

Dutfield, G., and Suthersanen, U. (2004) "The innovation dilemma: Intellectual property and the historical legacy of cumulative creativity" *Intellectual Property Quarterly* 8(3): 79–421.

Duysters, G., Kok, G., and Vaandrager, M. (2000) "Crafting successful strategic technology partnerships" *R&D Management* 29(4): 343–351.

Ebersole, T., Guthrie, M., and Goldstein, J. (2005) "Patent pools as a solution to the licensing problems of diagnostic genetics" *Intellectual Property & Technology Law Journal* 17(1): 6–13.

ECIS (European Commission Information Society) (2004) "Intellectual Property Rights and Digital Rights Management Systems Intellectual Property Rights which work in the digital era are essential" *Factsheet 20.*

Eden, M. (1984) "The engineering-industrial accord: Inventing the technology of health care" *Machine at the Bedside.* S. Reiser, and Anbar, M. Cambridge, Cambridge University Press: 49–64.

Edler, J., Blind, K., Frietsch, R., and Schmoch, U. (2004) *The scope and nature of the patent surge: A view from Germany.* Patents, Innovation and Economic Performance—2004 OECD Conference Proceedings (August 28–29, 2004, Paris, Committee for Science and Technology Policy, OECD.

EFF (2003) *Unintended consequences: Five years under the DMCA* (Version 3), EFF (Electronic Frontier Foundation).

Eisenberg, R. (1989) "Patents and the progress of science: Exclusive rights and experimental use" *The University of Chicago Law Review* 56: 1017–1086.

Eisenberg, R. (1996a) "Public research and private development: Patents and technology transfer in government-sponsored research" *Virginia Law Review* 82: 1663–1727.

Eisenberg, R. (1996b) *Patenting research tools and the law.* Intellectual Property Rights and Research Tools in Molecular Biology: Summary of a workshop held at the National Academy of Sciences, February 15–16, 1996, National Research Council.

Ellenbogen, J. (1997) *Matter as software.* Software Engineering and Economics Conference, March 1997, McLean, VA, MITRE Corporation.

Elliesen, T. (2006) "Das Ende der Freiheit" *Das Parlament* 10 (March 6, 2006).

Ellis, D. (2008) "Exxon shatters profit records", CNN.com, February 1, 2008

Encaoua, D., and Hollander, A. (2002) "Competition policy and innovation" *Oxford Review of Economic Policy* 18(1): 63–79.

Energy Economist (2005) "New wind turbine plant in India" *Energy Economist* (284): 32.

Engardio, P., and Einhorn, B. (2005) "Outsourcing innovation" *Business Week* (3925): 84.

Engdahl, W. (2004) "Gene-manipulated seeds: Are we losing our food security too?" *Current Concerns* 4.

Enomoto, A., Moore, M., Nakano, T., Egashira, R., Suda, T., Kayasuga, A., Kojima, H., Sakakibara, H., and Oiwa, K. (2006) *A molecular communication system using a network of cytoskeletal filaments.* 2006 NSTI Nanotechnology Conference.

Enos, J. (1962) *Petroleum progress and profits: A history of process innovation* Cambridge, MA, MIT Press.

EPO (2005) "Meeting the Nanotech challenge" *Intellectual asset management* (Supplement Patent in Europe): 14–15.

Epstein, R. (2005) "'Einspruch, Euer Ehren!'" *Technology Review* 7 (July 2005): 82–86.

Ernsberger, D. (2003) "US pushes for hydrogen-based economy" *Platt's Oilgram News* 81(32): 5.

ESA (2006) *Coexistence—The seed industry's views and expectations.* Co-existence of Genetically Modified, Conventional and Organic Crops, Vienna, European Seed Association.

ETC Group (2001) "New enclosures: Alternative mechanisms to enhance corporate monopoly and BioSerfdom in the 21st century" *ETC Group Communiqué* (73): 1–20.

ETC Group (2003) "Oligopoly, Inc.: Concentration in corporate power: 2003" *Communiqué* (82).

ETC Group (2005) "Nanotech's 'second nature' patents: Implications for the global south, ETC Group special report" *ETC Group Communiqué* (87 and 88).

Euler, R. (2006) "Kaum Interesse an Gen-Pflanzen" *Frankfurter Allgemeine Zeitung.* Frankfurt (14. August 2006).

Eurostat (2005) *Eurostat Jahrbuch 2005—Europa in Zahlen.* Luxemburg, European Commission.

Evans, L. (2003) "Challenges and opportunities in licensing to emerging economies" *LES Nouvelles* (December 2003): 163–166.

Evers, J. (2001) "Palm: Chip makers ready Palm OS for ARM chips" *Computer-World.com* (July 24, 2001)

Evers, J. (2006) "Intel's VPro to boost security" *CNET News.com* (April 24, 2006).

Fagerberg, J., Mowery, D., and Nelson, R., Ed. (2005) *The Oxford handbook of innovation.* Germany, Oxford University Press.

Fai, F., and von Zedtwitz, M. (2004) "Copyright gilt nur auf dem Papier" *IO New Management* (12): 46–50.

FAO (1999) *World list of seed sources*, United Nations Food and Agriculture Organisation.

FAO (2004) *FAO statistical yearbook*, United Nations Food and Agriculture Organisation.

FAZ (2006) "Rede zur Lage der Nation: Bush warnt vor Amerikas ‚Abhängigkeit vom Öl'. *Frankfurter Allgemeine Zeitung.* Frankfurt (February 1, 2006).

FDA (1990) "Device Recalls: A Study of Quality Problems" *HHS Publication 90–4235.* Rockville, Maryland, US Dept. of Health and Human Services, Public Health Service, Food and Drug Administration, Centre for Devices and Radiological Health.

Feller, J., and Fitzgerald, B. (2002) *Understanding open-source software development.* Boston, Addison Wesley.

Felten, E. (2004) "Time to Retire 'Hacking'", www.freedom-to-tinker.com/bbg/felten/time-retire-hacking – last accessed November 30, 2008

Felten, E. (2006) "DRM wars: Property rights management" www.freedom-to-tinker.com/?p=1052—last accessed April 14, 2007.

Ferguson, N. (2001) Censorship in action: Why I don't publish my HDCP results. (August 15, 2001) www.macfergus.com/niels/dmca/cia.html—last accessed June 12, 2007.

Fernandez Delicado, R., Bellver Cabello, D., and Lloro Boada, I. (2005) "The quantum cryptography: Communication and computation" *Acta Astronautica* 57(2–8): 348–355.

Fernandez–Cornejo, J. (2004) "The seed industry in US agriculture: An exploration of data and information on crop seed markets, regulation, industry structure, and research and development" *Agricultural Information Bulletin (Economic Research Service/USDA)* (AIB–786).

Fernandez–Cornejo, J., and Schimmelpfennig, D. (2004) "Have seed industry changes affected research effort?" *Amber Waves, USDA Economic Research Service.*

Financial Times (2006) "Oil spike gives BP room to return up to $65 bn" *Financial Times*: 1. (February 8, 2006).

Finkelstein, S., Neels, K., and Bell, G. (1995) "Innovation in cardiac imaging" *Sources of Medical Technology: Universities and Industry.* A. Gelijns, and Rosenberg, N., Institute of Medicine: 125–154.

Fisher, M. (2004) "Growth of the Japanese patent system: A lesson for us all?" *Intellectual Property Quarterly* (1): 85–113.

Focus (2006) "iPods sollten Sprengstoff zünden" *Focus* (August 11, 2006).

Foster, E. (2002) "Register, or else . . ." *InfoWorld* 24(40): 61.

Fowler, C., and Mooney, P. (1990) *Shattering: Food, politics, and the loss of genetic diversity.* Tucson, Arizona, University of Arizona Press.

Foxall, G., and Johnston, B., (1987) "Strategies of user-initiated product innovation" *Technovation* 6: 77–102.

Franke, N. (2003) "Toolkits for user innovation: die Einbindung des Kunden in den Innovationsprozess" *Die Gestaltung der Organisationsdynamik.* Stuttgart, Schäffer–Poeschel Verlag: 357–381.

Franke, N., and Piller, F. (2004) "Value creation by toolkits for user innovation and design: The case of the watch market" *Journal of Product Innovation Management* 21: 401–415.

Franke, N., and Schreier, M. (2002) "Entrepreneurial opportunities with toolkits for user innovation and design" *The International Journal on Media Management* 4(4): 225–235.

Franke, N., and Shah, S. (2003) "How communities support innovative activities: An exploration of assistance and sharing among end-users" *Research Policy* 32(1): 157–178.

Franke, N., and von Hippel, E. (2003) "Satisfying heterogeneous user needs via innovation toolkits: the case of Apache security software" *Research Policy* 32(7): 1199–1215.

Franke, N., von Hippel, E., and Schreier, M. (2006) "Finding commercially attractive user innovations: A test of lead user theory" *Journal of Product Innovation Management* 23: 301–315.

Franz, K. (1999) *Narrating automobility: Travellers, tinkerers, and technological authority in the twentieth century* Providence, R.I., Brown University, Ph.D. dissertation.

Freel, M. (2003) "Sectoral patterns of small firm innovation, networking and proximity" *Research Policy* 32(5): 751–770.

Freeman, C. (1968) "Chemical process plant: Innovation and the world market" *National Institute Economic Review*: 29–57.

Freiberger, P., and Swaine, M. (2000) *Fire in the valley: The making of the personal computer.* New York, McGraw–Hill.

Fritsch, M. and Slavtcher, V. (2005) "The Role of Regional Knowledge for Innovation", 45th ERSA Conference Amsterdam, August 23-27, 2005

Fritschel, H., Pandya-Lorch, R., and Rose, B. (1996) "Key trends in feeding the world, 1970–95" International Food Policy Research Institute.

Frossard, D. (1995) "Asia's green revolution and peasant distinctions: Between science and authority" http://lucy.ukc.ac.uk/Postmodern/David_Frossard_TOC.html—last accessed June 2, 2007.

Fuglie, K., Ballenger, N. and Day, K. (1996) *Agricultural research and development: Public and private investments under alternative markets and institutions*, AER-735, Economic Research Service, United States Department of Agriculture.

Galbraith, J. (1952) *American capitalism: The concept of countervailing power.* Cambridge, MA, Houghton Mifflin.

Galli, P. (2005) "Microsoft tightens piracy policy; Updates hinge upon software validation" *eweek* 22(30): 23.

Gallini, N. (1984) "Deterrence by market sharing: A strategic incentive for licensing" *American Economic Review* 74: 931–941.

Gallini, N., and Scotchmer, S. (2002) "Intellectual property: When is it the best incentive mechanism?" *Innovation policy and the economy.* A. Jaffe, Lerner, J. and Stern, S., MIT Press: 51–78.

Gallini, N., and Winter, R. (1985) "Licensing in the theory of innovation" *RAND Journal of Economics* 16: 237–252.

Gallini, N., and Wright, B. (1990) "Technology transfer under asymmetric information" *RAND Journal of Economics* 21: 147–160.

Gambardella, A. (1995) *Science and innovation: The US pharmaceutical industry during the 1980s.* Cambridge, Cambridge University Press.

Gambardella, A., Giuri, P., and Luzzi, A. (2006) *The market for patents in Europe*, Laboratory of Economics and Management, Sant'Anna School of Advanced Studies.

Gamesa (2004) *Gamesa annual report 2004*, http://www.gamesa.es/gamesa/modules/idealportal/uploadlink/ANNUALREPORT2004.pdf—last accessed May 16, 2007.

Gana, R. (1995) "Has creativity died in the third world? Some implications of the internationalization of intellectual property" *Denver Journal of International Law & Policy* 24(1): 109–144.

Gans, J., Hsu, D., and Stern, S. (2000) *When does start-up innovation spur the gale of creative destruction?* Cambridge, MA, NBER Working Paper No. 7851.

Garfinkel, S. (1995) "Patented secrecy" *Forbes* 155(5): 122–124.

Garner, B., Ed. (2004) *Black's law dictionary.* Chicago B&T.

Gartenberg, M. (2005) "Business must be cautious with Firefox" *Computerworld* 39(4): 21.

Gartner Dataquest (2003) *Linux on the desktop: The whole story.* Report No. AV-2–6574, Gartner Dataquest.

Gassmann, O. (2006) "Opening up the innovation process: towards an agenda" *R&D Management* 36(3): 223–228.

Gassmann, O., Sandmeier, P., and Wecht, C. (2006). "Extreme customer innovation in the front–end: Learning from a new software paradigm" *International Journal of Technology Management* 33(1): 46–67.

Gawer, A., and Cusumano, M. (2002) *Platform leadership.* Boston, MA, Harvard Business School Press.

Geiger, C., Kur, A., and Senftleben, M. (2003) "Report—A new framework for intellectual property rights—Conference of the Max Planck Institute for Intellectual Property at Elmau Castle, 22–23 November, 2003" *International Review of Industrial Property and Copyright Law* 34(6): 632–646.

Gelijns, A., and Halm, E., Ed. (1991) *The changing economics of medical technology.* Medical innovation at the crossroads: Vol. 2. Washington, D.C., National Academy Press.

Gelijns, A., and Rosenberg, N. (1995) *Sources of medical technology: Universities and industry*, Institute of Medicine, Washington D.C., National Academy Press.

Gelijns, A., and Rosenberg, N. (1999) "Diagnostic devices: An analysis of comparative advantages" *Sources of industrial leadership: Studies of seven industries.* D. Mowery, and Nelson, R. Cambridge, Cambridge University Press: 312–358.

Germain, J. (2004) "Identity theft online: Debunking the myths" *TechNewsWorld* (January 17, 2004).

Gibbs, J. (1968) "On crime, punishment, and deterrence" *Southwestern Social Science Quarterly* 49.

Gillespie, T. (2004) "Copyright and commerce: The DMCA, trusted systems, and the stabilization of distribution" *The Information Society* 20: 239–254.

Ginarte, J., and Park, W. (1997) "Determinants of patent rights: A cross–national study" *Research Policy* 26: 283–301.

Ginsburg, J. (2000) "From having copies to experiencing works: The development of an access right in US copyright law" *US Intellectual Property: Law and Policy.* Fordham H. Hansen.

Glaser, W. (1999) "Now, it's the motorcar" *Malaysian Business.* Kuala Lumpur: 54.

Glazier, S. (1997) *Patent strategies for business.* Washington, D.C., Law & Business Institute.

Glover, T. (2003) "Microsoft loses $10 billion a year from piracy", *Knight Ridder Tribune Business News*, Washington, April 6, 2003: 1

GNU (2007) "GNU general public license" www.gnu.org/copyleft/gpl.html—last accessed February 19, 2007.
Goedkoop, M., Halen, C., and de Riele, H. (1999) *Product service systems*, Dutch Ministries of Environment, Dutch Ministries of Economic Affairs.
Golden, C. (2002) "Toner cartridge computer chip usage and the impact on the aftermarket" *Recharger Magazine* (December 2002): 36–46.
Goldstein, S. (2001) "Cloud of DVD licensing fees hangs over replicators" *Video Store Magazine* 23(24): 14.
Goral, T. (2003) "Recording industry goes after campus p-2-p networks: suit alleges $97.8 billion in damages" *University Business* May 2003.
Graaf, B., Lormans, M. and Toetenel, H. (2003) "Embedded software engineering: The state of the practice" *IEEE Software* (November/December 2003): 61–69.
Graham, S., and Mowery, D. (2004) "Submarines in software? Continuation patenting in software in the 1980s and 1990s and its implications for open source" *Economics of Innovation and New Technology* 13(5): 443–456.
GRAIN (2000) "Hybrid rice in Asia: An unfolding threat" *GRAIN*: 1–24.
GRAIN (2004) "Interview: Hope shand" *Seedling* (April 2004): 22–24.
GRAIN (2005a) "Seed laws: Imposing agricultural apartheid" *Seedling* (July 2005).
GRAIN (2005b) *Iraq's new patent law: A declaration of war against farmers*. London Focus on the Global South and GRAIN.
Granstrand, O. (1999) *The economics and management of intellectual property: towards intellectual capitalism*. Cheltenham, U.K:, Edward Elgar.
Greene, T. (2001) "MPAA believes all Netizens are criminals" *The Register* (April 5, 2001).
Greenberg, A. (2007) "Broadband Big Brothers', *Forbes* (October 12, 2007)
Greenleaf, W. (1961) *Monopoly on wheels: Henry Ford and the Selden automobile patent*. Detroit, Wayne State University Press.
Greenpeace (2007) Say no to genetic engineering, Greenpeace Campaign, Greenpeace, www.greenpeace.org/international/campaigns/genetic–engineering—last accessed February 20, 2007.
Greenstein, S. (1997) "Lock-in and the costs of switching mainframe computer vendors: What do buyers see?" *Industrial and Corporate Change* 6(2): 247–273.
Greipl, E., and Täger, U. (1982) *Wettbewerbswirkungen der unternehmerischen Patent- und Lizenzpolitik*. Berlin, Duncker & Humblot.
Griliches, Z. (1964) "Research costs and social returns: Hybrid corn and related innovations" *Agriculture in Economic Development*. C. Eicher. New York, McGraw-Hill: 369–386.
Grimm, T. (2004) *User's guide to rapid prototyping*. Dearborn, Michigan, Society of Manufacturing Engineers.
Grindley, P., and Teece, D. (1997) "Managing intellectual capital: Licensing and cross-licensing in semiconductors and electronics" *California Management Review* 39: 1–34.
Guckelsberger, U. (2005) *Das Menschenbild in der Ökonomie*, FH Ludwigshafen, http://web.fh–ludwigshafen.de/fb2/guckelsberger.nsf/Files/1D61686B4BEEC666C1256F9E00412045/$FILE/Der%20Homo%20oeconomicus.doc—last accessed June 12, 2006.
Guinnessy, P. (1997) "Tooth delay banished" *New Scientist* (2090) (July 12, 1997).
Gupta, R., and Wilemon, D. (1996) "Changing patterns in industrial R&D management" *Journal of Product Innovation Management* 13(6): 497–511.
Hagedoorn, J. (2002) "Inter-firm R&D partnerships: An overview of major trends and patterns since 1960" *Research Policy* 31(4): 477–492.
Hall, B. (2004) *Exploring the patent explosion*. NBER working paper series: 10605. Cambridge, MA.

Hall, B., and Ziedonis, R. (2001) "The patent paradox revisited: An empirical study of patenting in the US semiconductor industry, 1979–1995" *RAND Journal of Economics* 32(1): 101–128.

Hamel, G., and Prahalad, C., (1990) "The core competence of the corporation" *Harvard Business Review* 90(3): 79–90.

Hamel, G. and Prahalad, C., (1994) *Competing for the Future* Boston, MA, Harvard Business School Press

Hamman, P., and Mittag, H. (1986) "The marketing of industrial technology through licensing" *Industrial marketing*. K. Backhaus, and Wilson, D. Berlin, Springer.

Hantler, S. (2004) "The seven myths of highly effective plaintiffs' Lawyers" *Civil Justice Forum* 42(April 2004).

Hantman, R. (1985) "Experimental use as an exception to patent infringement" *Journal of the US Patent and Trademark Office* 65.

Harhoff, D., Henkel, J., and von Hippel, E. (2003) "Profiting from voluntary information spillovers: How users benefit from freely revealing their innovations" *Research Policy* 32(10): 1753–1769.

Harish, I. (2004) "Wilson to market nanotech based sports equipment" *Businessline*. Chennai, (October 10, 2004): 1.

Harmon, R. (2001) "Injunctive relief against an infringer is the norm [in decisions by the CAFC]" *Patents and federal circuit*. R. Harmon. Washington, Bureau of National Affairs.

Harper, B., and Udupa, V. (2004) "Drafting electronic software licenses to prevent reverse engineering" *E-Commerce Law Report* 6(2): 18.

Harrigan, K. (1985) "Exit Barriers and Vertical Integration", *Academy of Management Journal* 28(3): 686-697

Harris, A., and Yen, D. (2002) "Biometric authentication: Assuring access to information" *Information Management & Computer Security* 10(1): 12–19.

Harris, M., and Massey, D. (1968) *Vertical coordination via contract farming*. Washington, D.C., US Department of Agriculture.

Harrison, I. (2004) *The Book of Inventions—The stories behind the inventions and inventors of the modern world*. London, Cassel Illustrated.

Harvey, A., Speksnijder, G., Baugh, L., Morris, J. and Ivarie, R. (2002) "Consistent Production of Transgenic Chickens Using Replication-deficient Retroviral Vectors and High-throughput Screening Procedues", Poultry Science 81: 202-212

Hayes, R., and Wheelwright, S. (1984) *Restoring our competitive edge: Competing through manufacturing*. New York, John Wiley.

Hearn, P. (1986) *The business of industrial licensing*, 2nd ed. Aldershot, England, Gower.

Hebert, P. (2005) *Nanotechnology gold rush yields crowded, entangled patents*. New York, Lux Research and Foley & Lardner LLP.

Heller, M., and Eisenberg, R. (1998) "Can patents deter innovation? The anticommons in biomedical research" *Science* 280: 698–701.

Henkel, J. (2004) "Patterns of free revealing—Balancing code sharing and protection in commercial open source development" http://opensource.mit.edu/papers/henkel2.pdf (August 2004).

Henkel, J. (2006) "Selective revealing in open innovation processes: The case of embedded linux" *Research Policy* 35(7) 953–969

Henkel, J., and von Hippel, E. (2005) "Welfare implications of user innovation" *Journal of Technology Transfer* 30(1–2): 73–87.

Herbst, S., and Volling, A. (2006) "Monsanto, Verträge und abhängige Bauern" *Bauernstimme* June 2006.

Herndon, J. (2002) "Intellectual property, antitrust, and the economics of aftermarkets" *Antitrust Bulletin* 47(2–3): 309–335.

Herstatt, C. (1991) *Anwender als Quellen für Produktinnovation*. Rechts- und staatswissenschaftliche Fakultät. Zurich, Universität Zürich.

Herstatt, C. (2004) "Trendsetter liefern die Idée" *IO New Management* (6): 18–24.
Herstatt, C., and von Hippel, E. (1992) "From experience: Developing new product concepts via the lead user method" *Journal of Product Innovation Management* 9(3): 213–222.
Herstatt, C., and Lettl, C. (2004) "Management of 'technology push' development projects" *International Journal of Technology Management,* 27(2–3): 155–175.
Hesseldahl, A. (2006) "A quantum leap in data encryption" *Business Week,* (November 6, 2006).
Hienerth, C. (2006) "The commercialization of user-innovations: The development of the rodeo kayak industry" *R&D Management* 36(3): 273–294.
Hitt, G. (2005) "US to pressure China at WTO on piracy fight" *Wall Street Journal* (October 26, 2005): A.17.
Hoff, P. (1986) *Inventions in the market place—Patent licensing and US antitrust laws.* Washington, American Enterprise Institute for Public Policy.
Holt, K. (1988) "The role of the user in product innovation" *Technovation* 7: 249–258.
Homiller, D. (2005) *The Digital Millennium Copyright Act and the European Union Copyright Directive: Next steps* Duke University School of Law, Duke University.
Hope, J. (2004) *Open–source biotechnology,* Canberra, The Australian National University.
Horgan, J. (1998) *The end of science.* London, Abacus.
Hossain, S., and Islam, M. (1985) *A report on the first workshop of innovative rice farmers.* Mymensingh, Bangladesh Agricultural University.
Howells, J. (1999) "Research and technology outsourcing" *Technology Analysis & Strategic Management* 11(1): 17–29.
Huang, Z., Chen, H., Chen, Z–K., and Roco, M. (2004) "Perspectives international nanotechnology development in 2003: Country, technology field analysis based on USPTO patent database" *Journal of Nanoparticle Research* 6: 325–354.
Hugenhaltz, B. (2000) "Why the Copyright Directive is Unimportant and Possibly Invalid", European Intellectual Property Review 22(11): 499-505
Hulme, G. (2003a) "The mind of a hacker" *InformationWeek* (963): 42.
Hulme, G. (2003b) "Flurry of arrests made in Cybercrime cases" *InformationWeek* (955): 22.
Huppertz, M.-T. (2004) *The role of IPR for the software industry: A changing landscape.* Patents, Innovation and Economic Performance—2004 OECD Conference Proceedings 28 & 29th August, Paris, Committee for Science and Technology Policy, OECD.
Hutton, J. (1991) "Medical device innovation and public policy in the European economic community" *The changing economics of medical technology.* A. Gelijns, and Halm, E. Washington, D.C, National Academy Press: 141–154.
IBM (2004) *IBM annual report 2004.* www.ibm.com.
IBM (2006) "First–of–a–kind (FOAK) program" *IBM,* http://www-306.ibm.com/software/success/cssdb.nsf/CS/LJKS-6RNQ4G?OpenDocument&Site—last accessed March 3, 2006.
Ikerd, J. (2000) *New farmers for a new century.* Youth in Agriculture Conference, Ulvik, Norway (February 14–15, 2000).
ILO (2004) *World employment report 2004/2005* Geneva International Labour Organization.
IRI (2007) "Industrial Research Institute R&D Trends Forecast for 2007", Research Technology Management, January/February: 17-20
Irle, M. (2005) "'Spreche ich eigentlich kein Deutsch mehr?'" *Brand eins* 6/2005.
Jackson, D. (2001) "Settling the flap surrounding WAP" *Telephony* 240(2): 28.
Jacobson, T., and Smith, G. (2001) *Cotton's renaissance—A study in market innovation.* Cambridge, Cambridge University Press.
Jaffe, A., and Lerner, J. (2004) *Innovation and its discontents: How our broken patent system is endangering innovation and what to do about it.* Princeton, Princeton University Press.

Janaiah, A. (2005) Hybrid rice: The Indian experience and comparison with the rest of Asia. *India development report 2004–5*. New Delhi, Oxford University Press: 141–153.

Jänicke, M., and Jacob, K. (2004) "Lead markets for environmental innovations: A new role for the nation state" *Global Environmental Politics* 4(1): 29–37.

Jarvie, N. (2003a) "Control of cybercrime—Is an end to our privacy on the internet a price worth paying?" *Computer and Telecommunications Law Review* 9(3): 76–81.

Jarvie, N. (2003b) "Control of cybercrime—Is an end to our privacy on the internet a price worth paying? Part 2" *Computer and Telecommunications Law Review* 9(4): 110–115.

Jewkes, J., Sawers, D., and Stillerman, R. (1969) *The sources of invention*. London, Macmillan.

John Deere (2007) Crop tracer, John Deere, http://www.deere.com/en_ZA/products/ag/ams/croptracer.html—last accessed February 24, 2007.

Johnson, C., Macleod, R., Parker, S., and Weinstein, D. (2004) "Biomedical computing and visualization software environments" *Association for Computing Machinery. Communications of the ACM* 47(11): 64–72.

Johnson, S. (2001) "Marketplace of ideas or tag sale?" *The Nation* 273(20): 25–28.

Johnson, C. (2006) "Encryption technology makes quantum leap", *EE Times* (October 2, 2006)

Johnson-Laird, A. (1994) "Software reverse engineering in the real world" *University of Dayton Law Review* 19: 843.

Jones, B. (2005) *The burden of knowledge and the 'death of the renaissance man': Is innovation getting harder?* NBER working paper No. 11360.

Jorash, P. (2006) *Genetic use restriction technologies–GURTs*. Bonn, Bundesverband Deutscher Planzenzüchter.

Jorgensen, U. (1997) *Danish energy and technology policies with respect to alternative energy sources and innovation*. tip Workshop: Regulation and Innovative Activities: Electricity Supply Industry, Vienna.

Joy, B. (2000) "Why the future does not need us" *Wired* (8.04).

Kahn, A. (1991) "The dynamics of medical device innovation: An innovator's perspective" *The changing economics of medical technology*. A. Gelijns, and Halm, E. Washington, D.C, National Academy Press: 89–95.

Kaiser, R. (2002) *Technological paradigm shifts and new modes of coordination in science-based industries*. Fourth Triple Helix Conference "Breaking boundaries—Building bridges" November 6–9, 2002, Copenhagen/Lund.

Kaiser, U., and Ronde, T. (2005) "A Danish view on software-related patents" *Centre for Economic and Business Research (CEBR)*: 1–20.

Kasarda, J., and Rondinelli, D. (1998) "Innovative infrastructure for agile manufacturers" *Sloan Management Review* 39(2): 73–82.

Kästler, G. (2005) "Europe's seed laws: locking out farmers" *Seedling* (July 2005): 10–16.

Kastriner, L. (1991) "The revival of confidence in the patent system" *Journal of the Patent and Trademark Office Society* 73(January): 5–23.

Katz, M., and Shapiro, C. (1985) "On the licensing of innovations" *RAND Journal of Economics* 16: 504–520.

Katz, R., Becker, R., Hayes, S., and Hanse, S. (2004) "Patents give the DMCA needed teeth: Salvaging copy protection in the post-Grokster era" *Journal of Internet Law* 8(3): 15–17.

Keegan, P. (2001a) "Lego: Intellectual property is not a toy" *Business 2.0* (September 2001).

Keegan, P. (2001b) "Go forth and hack" *Business 2.0* (November 2001).

Keeley, L. (2005) "Innovation vs. Design" cph127, http://www.cph127.com/cph127/2005/11/innovation_vs_d.html—last accessed April 5,2007.

Kemmling, C. (2006) "Das höchste der Gefühle" *Yacht*, 21: 42–49.
Kerr, D. (1996) "Public key mystery" *Computerworld* 30 (37): 93–94.
Kerr, I., Maurushat, A., and Tacit, C. (2003) "Technical protection measures: Part II The legal protection of TPMs" Ottawa Nelligan O'Brien Payne.
Kesan, J. (2000) "Intellectual property protection and agricultural biotechnology" *The American Behavioral Scientist* 44(3): 464–503.
Kesan, J., and Janis, M. (2002) "Intellectual property protection for plant innovation: Unresolved issues after J.E.M v. Pioneer" *Nature Biotechnology* 20: 1161–1164.
Khoury, S., and Sullivan, J. (2005) *How to make money out of your IP—The importance of licensing*, LESI World Conference, Munich (June 15, 2005).
Kingston, W. (2000) *Enforcing small firms' patent rights.* Dublin CORDIS.
Kingston, W. (2004) "Making patents useful to small firms" *Intellectual Property Quarterly* (4): 369–78.
Klein, B. (1996) "Market power in aftermarkets" *Managerial and Decision Economics* 17(2): 143–164.
Klein, B. (1997) "Why hold-ups occur: The self-enforcing range of contractual relationships" *Uncertainty and Economic Evolution: Essays in Honour of Armen A. Alchian.* London, Routledge: 63–86.
Klemperer, P. (1987) "Markets with consuming switching costs" *Quarterly Journal of Economics* 102: 375–394.
Kline, R., and Pinch, T. (1994) *Taking the black box off its wheels : The social construction of the car in the rural United States.* Berlin, WZB Wissenschaftszentrum Berlin für Sozialforschung.
Knapp, L. (2003) "Printer fails after run-in with generic ink cartridge" *The Seattle Times.* Seattle (May 17, 2003).
Knight, K. (1963) *A study of technological innovation: The evolution of digital computers.* Pittsburgh, PA Carnegie Institute of Technology.
Ko, Y. (1992) "An economic analysis of biotechnology patent protection" *The Yale Law Journal* 102: 777–804.
Kodama, F. (1995) *Emerging patterns of innovation* Boston, MA, Harvard Business School Press.
Koenig, C., Neumann, A. and Katzschimann, T. (2004) *Trusted Computing,* Heidelberg, Verlag Recht und Wirtschaft
Koepf, H., and von Plato, B. (2001) *Die biologisch-dynamische Wirtschaftsweise im 20. Jahrhundert.* Dornach, Verlag am Goetheanum.
Kollmer, H., and Dowling, M. (2004) "Licensing as a commercialisation strategy for new technology–based firms" *Research Policy* 33: 1141–1151.
Kortum, S. (1993) "Equilibrium R&D and the decline in the patent-R&D ratio: U.S. evidence", *American Economic Review Papers and Proceedings*, May, 450-457
Kortum, S., and Lerner, J. (1997) *Stronger protection or technological revolution: What's behind the recent surge in patenting?* Working paper 98–012. Boston, MA, Harvard Business School.
Kortum, S., and Lerner, J. (1998) *Does venture capital spur innovation?* Working paper No. 6846, Cambridge, MA, National Bureau of Economic Research.
Krause, J. (2005) "Computing time for a crime" *ABA Journal* 91: 14.
Krebs, A. (2004) "New Iraq patent law will make traditional farmers seed saving illegal" *The Agribusiness Examiner* (380).
Kremen, S. (1998) "Apprehending the computer hacker: The collection and use of evidence" *Computer Forensics Online*, http://www.shk–dplc.com/cfo/articles/hack.htm -last accessed May 23, 1998.
Krugman, P. (1991) *Geography and trade.* Cambridge, MA, The MIT Press.
Krummenacker, M. "Problems with the current intellectual pseudo "property" (IPP) system, specifically with patents" Article prepared for the Foresight Institute

Spring 2000 Senior Associate Gathering Confronting Singularity, www.n-a-n-o.com/ipr/fi-gathering2/ipp-scenarios.html—last accessed on February 13, 2007.

Kuhlmann, D. (2004) "Open trusted computing technopolihyche herausforderung", *Trusted Computing*, Koenig, C. Neumann, A. and Katzschmann, T., Heidelberg, Verlag Recht und Wirtschaft

Kumar, V., Kumar, U., and Persaud, A. (1999) "Building technological capability through importing technology: The case of the Indonesian manufacturing industry" *The Journal of Technology Transfer* 24(1): 81–96.

Kurzweil, R. (1999) *The age of spiritual machines* New York, Penguin Books.

LaMonica, M. (2008) "Eco-patent commons shares earth-friendly tech" *CNET News.com* (January 13, 2008).

Labrador, D. (2002) „Teaching robot dogs new tricks" *Scientific American* (January 21, 2002).

Lacy, W. (2000) "Commercialization of university research brings benefits, raises issues and concerns" *California Agriculture* 54(4): 72–79.

Lakhani, K., and von Hippel, E. (2003) "How open source software works: Free user-to-user assistance" *Research Policy* 32(6): 923–943.

Lambke, A., Janßen, G., and Schievelbein, C. (2003) "Der Streit ums Saatgut" *Der Kritische Agrarbericht 2003*: 70–78.

Lammerts van Bueren, E. (2002) *Organic plant breeding and propagation: Concepts and strategies.* Driebergen, The Netherlands, Louis Bolk Institute.

Lamoreaux, N., and Sokoloff, K. (1997) *Inventors, firms, and the market for technology: US manufacturing in the late nineteenth and early twentieth century.* Historical paper 98. Cambridge, MA, National Bureau of Economic Research.

Lanjouw, J., and Lerner, J. (1997) *The enforcement of intellectual property rights: A survey of the empirical literature.* Working paper w.8656. Cambridge, MA, National Bureau of Economic Research.

Lanjouw, J., and Schankerman, M. (2004) "Protecting intellectual property rights: Are smaller firms handicapped?" *Journal of Law and Economics* 47(1): 45–74.

Laube, H. (2006) "Intel erringt Teilsieg gegen AMD" *Financial Times Deutschland* (September 28, 2006).

Le, M. (2005) "'Terminator' technology keeps GM crops in check" *New Scientist* (February 26, 2005).

Le, N. (2004) "Microsoft Europe and switching cost" *World Competition: Law and Economics Review* 27(4): 567–594.

Ledwith, A., and Coughlan, P. (2005) "Splendid isolation: Does networking really increase new product success?" *Creativity and Innovation Management* 14(4): 366–373.

Lee, S., Moisa, N., and Weiss, M. (2003) "Open source as a signalling device—An economic analysis" opensource.mit.edu/papers/leemoisaweiss.pdf (March 21, 2003).

Leiby, P., and Rubin, J. (2004) "Understanding the transition to new fuels and vehicles: Lessons learned from analysis and experience of alternative fuel and hybrid vehicles" *The Hydrogen Energy Transition-Moving towards the post petroleum age in transportation.* D. Sperling, and Cannon, J., Burlington, MA 191–212.

Leifer, R., McDermott, C., O'Connor, G., Peters, L., Rice, M. and Veryzer, R. (2000) *Radical innovation how mature companies can outsmart upstarts.* Boston, MA, Harvard Business School Press.

Lemley, M. (2002) "Intellectual property rights and standard-setting organizations" *California Law Review* 90: 1889–1973.

Lemley, M. (2005) *Patenting nanotechnology.* Stanford, Law Review & 58(2): 601-630.

Lemos, R. (2001)"Security workers: Copyright law stifles" *CNET News* (September 6, 2001).

Leonard, D. (1998) *Wellsprings of knowledge.* Boston, MA, Harvard Business School Press.

Leonard, R. (2002) "Will history repeat itself? Challenges of the new high-tech market" *Recharger Magazine* (December 2002): 81–95.
LER (2001) "Interview with Emmet J. Murtha" *Licensing Economics Review* October 2001.
Lehman, J. (2006) "Intellectual property rights and Chinese tradition section: Philosophical foundations" *Journal of Business Ethics* 69(1): 1
Lerner, J. (1995) „Patenting in the shadow of competitors" *Journal of Law and Economics* 38(2): 463–495.
Lerner, J. (2002) *Patent protection and innovation over 150 years*. Cambridge, MA, National Bureau of Economic Research.
Lerner, J. (2004) *The new new financial thing: The sources of innovation before and after State Street*. Cambridge, MA, National Bureau of Economic Research, Working paper 10223: 1–54.
Lerner, J., and Tirole, J. (2002) "Some simple economics of open source" *Journal of Industrial Economics* 50(2): 197–232.
Lessig, L. (2004) *Free culture*. New York, The Penguin Press.
Lessig, L. (2001) *The future of ideas*. New York, Random House.
Lessig, L. (1999) *Code and other laws of cyberspace*. New York, Basic Books.
Lettice, J. (2002) "European antitrust chief concerned over MS Palladium?" *The Register* (July 2, 2002).
Lettl, C., Herstatt, C., and Gemuenden, H. (2006) "User's contributions to radical innovation: Evidence from four cases in the field of medical equipment technology" *R&D Management* 36(3): 251–272.
Levin, R., Klevorick, A., Nelson, R. and Winter, S. (1987) "Appropriating the returns from industrial research and development" *Brookings Papers on Economic Activity* 3: 783–820.
Levy, S. (1984) *Hackers: heroes of the computer revolution*. London, Penguin.
Levy, S. (2002) "Lawrence Lessig's supreme showdown" *Wired* (October 10, 2002).
Libin, K., and Gray, J. (2003) "RIM's patent wars" *Canadian Business* 76 (September 2, 2003).
Library of Congress (2002) "Exemption to prohibition on circumvention of copyright protection systems for access control technologies" *Federal Register* 67(199): 63578–63582.
Lie, D. (2003) "Architectural support for copy and tamper-resistant software" *Department of Electrical Engineering*. Stanford University, Stanford.
Lilien, G., Morrison, P., Searls, K., Sonnack, M., von Hippel, E., (2002) "Performance assessment of the lead user generation process for new product development" *Management Science* 48(8): 1042–1059.
Lim, K. (2000) *The many faces of absorptive capacity: spillovers of copper interconnect technology for semiconductor chips*. Working paper 4110. MIT Sloan School of Management, Cambridge, MA, MIT.
Lim, K. (2004) "The many faces of absorptive capacity: Spillovers of copper interconnect technology for semiconductor chips" Singapore, National University of Singapore.
Lin, J. (1992) "Hybrid rice innovation in China: A study of market-demand induced technological innovation in a centrally-planned economy" *The Review of Economics and Statistics* 74(1).
Linden, G., and Somaya, D. (2000) *System-on-a-chip integration in the semiconductor industry: Industry structure and firm strategies*. Working paper, Haas School of Business, Berkeley.
Linowitz, S. (1957) "Antitrust aspects of Grant Back clauses in license agreements" *Cornell Law Quarterly* 43: 217.
Litman, J. (2001) *Digital copyright*. Amherst, NY, Prometheus Books.
Lloyd, I. (2000) *Information technology law*. London, Butterworths.

Lockton, D. (2005) "Architectures of control in consumer product design" *Judge Institute of Management*. Cambridge, University of Cambridge.

Long, P. (2001) *Openness, secrecy, authorship*. Baltimore, The Johns Hopkins University Press.

Longping, Y. (2004) *Hybrid rice for food security in the world*. FAO Rice Conference, Rome, Italy.

Lotz, P. (1992) *Demand side effects on product innovation: The case of medical devices*. Copenhagen, Copenhagen Business School.

Louie, L. (2001) "A prescription for profits" *Upside* 13(5): 102–107.

Louwaars, N. (2005) "Biases and bottlenecks: Time to reform the South's inherited seed laws?" *Seedling* (July 2005): 4–9.

Love, J., and Roper, S. (2001) "Location and network effects on innovation success: Evidence for UK, German and Irish manufacturing plants" *Research Policy* 30: 643–661.

Lovell, E. (1969) *Appraising foreign licensing performance*. New York, National Industrial Conference Board.

Löwer, C. (2005) "Forscher imitieren Insektenaugen" *Wirtschafts Woche* (November 14, 2005).

Lu, M., Xin, Z., Cheng, S., He, Z., Li, X., Han, T., Yu, S., and Wang, H. (2004) *Recent advances in crop improvement in China*. New Directions for a Diverse Planet: Proceedings of the 4th International Crop Science Congress, Brisbane, Australia.

Lundquist, G. (2005) *Innovating technology transfer: Four levels of mindsets and processes*. Advancing Innovation and Entrepreneurship, Kansas City Kauffman Foundation.

Lüthje, C. (1999) *Kundenorientierung im Innovationsprozeß: Eine Untersuchung zur Customers-Hersteller-Interaktion auf Konsumgütermärkten*. Munich, Ludwig-Maximillians-Universität München.

Lüthje, C. (2003) *Customers as co-inventors: An empirical analysis of the antecedents of customer-driven innovations in the field of medical equipment*. Proceedings of the 32nd EMAC Conference, Glasgow.

Lüthje, C., and Herstatt, C. (2004) "The lead user method: An outline of empirical findings and issues for future research" *R & D Management* 34(5): 553–568.

Lüthje, C., Herstatt, C., and von Hippel, E. (2005) "The dominant role of "local" information in user innovation: The case of mountain biking" *Research Policy* 34(6): 951–965.

Lutzker, A., and Lutzker, S.J. (2005) "Altering the contours of copyright—The DMCA and the unanswered questions of Paramount Pictures Corp. v. 321 Studios" *Santa Clara Computer and High–Technology Law Journal* 21(3): 561–590.

Lux Research (2007) *The nanotech report*, 5th Ed, New York, Lux Research.

Lyman, J. (2005) "Bnetd reverse engineering ruling may stifle innovation" *NewsForge* (October 17, 2005).

Lyra Research (2005) *Think ink: The 2005 US ink jet cartridge user survey*. Newtonville, MA, Lyra Research.

MacDonald, J., and Korb, P. (2006) "Agricultural contracting update: Contracts in 2003" *USDA-ERS Economic Information Bulletin* (9).

MacDonald, J., Perry, J., Ahearn, M., Banker, D., Chambers, W., Dimitri, C., Key, N., Nelson, K., and Southard, L. (2004) *Contracts, markets, and prices: Organizing the production and use of agricultural commodities*. Washington, D.C. Economic Research Service, United States Department of Agriculture.

Machlup, F., and Penrose, E. (1950) "The patent controversy in the nineteenth century" *Journal of Economic History* 10(1): 24.

Macho-Stadler, I., Martinez-Giralt, X., and Pérez-Castrillo, J. (1996) "The role of information in licensing contract design" *Research Policy* 25: 43–57.

Macleod, C. (1988) *Inventing the Industrial Revolution* Cambridge, Cambridge University Press.
Maebius, S. (2004) "Ten patents that could impact the development of nanotechnology" *Nanotechnology Law & Business Journal* 1(2, 12): 1–7.
Mahdjoubi, D. (2004) *Knowledge, innovation and entrepreneurship: Business plans, capital, technology and growth of new ventures in Austin*, Texas. Austin, The University of Texas at Austin.
Maier, A. (2006) "Original und Fälschung" *enable—case studies*. F. T. Deutschland, Financial Times Deutschland: 24–27.
Maloney, L. (2005) "It's a small world after all" *Design News* (September 26, 2005).
Mamudi, S. (2005) "Washington's IP agenda" *Managing Intellectual Property* (April 2005): 17–21.
Manners, D. (2005) "Legal ruling shock hits NAND firms" *Electronics Weekly* (2190): 3.
Mansfield, E., Romeo, A., Schwartz, M., Teece, D., Wagner, S., and Brach, P. (1982) Technology Transfer, Productivity and Economic Policy. New York, W.W. Norton
Mansfield, E. (1986) "Patents and innovation: An empirical study" *Management Science* 32(2): 173–181.
Mansfield, E (1985) "How rapidly does new industrial technology leak out?" *The Journal of Industrial Economics* 34(2): 217.
Maromonte, K. (1998) *Corporate strategic business sourcing*. Connecticut, Quorum Books.
Marsh, S. (2004) "Peer Pressure" *Cable & Satellite Europe* (237): 1.
Martinez, C., and Guellec, D. (2004) *Overview of recent changes and comparison of patent regimes in the United States, Japan and Europe*. Patents, Innovation and Economic Performance—2004 OECD Conference Proceedings August 28–29, 2004, Paris, Committee for Science and Technology Policy, OECD.
Martinez, C., Sheehan, J. and Guellec, D. (2004) *Understanding business patenting and licensing: Results of a survey*. Patents, Innovation and Economic Performance—2004 OECD Conference Proceedings (August 28–29, 2004, Paris, Committee for Science and Technology Policy, OECD.
MASIPAG (2003) *Selling food.health.hope*, Los Baños MASIPAG, RESIST and PRESSURE POINT.
Matsunaga, Y. (1977) *Successful licensing to and from Japan*. Tokyo, Nihon Brain Corporation.
Maurer, S., and Scotchmer, S. (1999) "Database protection: Is it broken and should we fix it?" *Science* 284: 1129–1130.
Maurer, S., Rai, A., Sali. A. (2004) "Finding cures for tropical diseases: Is open source an answer?" *Biotechnology: Essays from its heartland*. L. Yarris, Berkley, CA Lawrence Berkeley Laboratory: 33–37.
Maurya, D. (1989) "The innovative approach of Indian farmers" *Farmer first—Farmer innovation and agricultural research*. R. Chambers, Pacey, A., and Thrupp, L. London, Intermediate Technology Publications: 9–14.
Maurya, D., Bottrall, A., and Farrington, J. (1988) "Improved livelihoods, genetic diversity and farmer participation: A strategy for rice-breeding in rainfed areas of India" *Experimental Agriculture* 24(3): 311–320.
Maxeiner, J. (2003) "Standard-terms contracting in the global electronic age: European alternatives" *Yale Journal of International Law* 28: 109–182.
McCullagh, D. (2002) "Apple: Burn DVDs—And we'll burn you" *CNET News.com* (August 28, 2002).
McCullagh, D. (2005) "Blizzard wins lawsuit on video game hacking" *CNET News.com* (September 2, 2005).

McDermott, C. M., and O'Connor, G.C. (2002) "Managing radical innovation: An overview of emergent strategy issues" *Journal of Product Innovation Management* 19: 424–438.

McGavock, D., and Lasinski, M. (1998) "IP survey finds 'gap' in information" *LES Nouvelles* (September 1998): 107–116.

McGuire, P. (1988) *The impact of product liability.* Conference Board Research Report No. 908.

McNutt, B., and Rodgers, D. (2004) "Lessons learned from 15 years of alternative fuels experience" *The Hydrogen energy transition-moving towards the post petroleum age in transportation.* D. Sperling, and Cannon, J.S., Burlington, MA, Elsevier Academic Press: 165–180.

Meek, J. (2000) "Genetic chickens get DNA copyright tag" *The Guardian.* London (July 31, 2000): 4.

Mello Jr., J. (2005) "Bill would force car makers to share computer code" *TechNewsWorld* (May 5, 2005).

Menzel, M., and Tauber, A. (2006) "Reisanbieter fürchten Imageschaden" *Financial Times Deutschland* (September 13, 2006).

Merck (2004) *Merck annual report 2004* http://www.merck.com/finance/annualreport/ar2004/chairmans_message—last accessed March 6, 2008.

Merges, R., and Nelson, R. (1990) "On the complex economics of patent scope" *Columbia Law Review* 90(4): 839.

Merritt, R. (2005) "Engineer steers Fox group's adventures in digital media." *Electronic Engineering Times* (1384): 1–3.

Meyer, M., and Lopez, L. (1995) "Technology strategy in a software products company" *Journal of Product Innovation Management* 12(4): 294–306.

Michael, D., and Rivette, K. (2004). *Facing the China challenge—Using an intellectual property strategy to capture global advantage*, Boston, MA, BCG Report.

Michaels, P. (2004) "Activation Nation" *Macworld* 21(6): 20–21.

Michaud, F., and Arpin, J. (2005) "United states grapples with proposed patent reform legislation" *Baker Bott's Intellectual Property Report* 5(53).

Microsoft decision (1999) United States of America Versus Microsoft Corporation, Civil Action No. 98-1232 (TP) "Findings of Fact", United States District Court for the District of Columbia 5.November 1999.

Microsoft (2006) "Original Microsoft Software" www.microsoft.com/genuine/downloads/WhyValidate.aspx—last accessed September 15, 2006.

Miller, J., Serrato, R., Represas-Cardenas, J., and Kundahl, G. (2005) *The handbook of nanotechnology: Business, policy, and intellectual property law.* Hoboken, NJ, John Wiley & Sons.

Millson, M., Raj, S., and Wilemon, D. (1996) "Strategic partnering for developing new products" *Research Technology Management* 39(3): 41–49.

Minoli, D. (2005) "Start planning for the nanorevolution" *Network World* 22(29): 41.

Minot, N. (1986) *Contract farming and its effect on small farmers in less developed countries, Department of Agricultural Economics.* Michigan State University.

Mishina, K. (1989) *Essays on technological evolution.* Cambridge, Harvard University.

Mollick, E. (2004) *Innovations from the underground: Towards a theory of parasitic innovation.* Cambridge, MIT: 133.

Monitor (2005) "Der Sektor Landwirtschaft" *Monitor Journal,* Nr. 4, December 2005.

Monsanto (2005) "Seeding values" *2005 Monsanto pledge report.* www.monsanto.com/monsanto/content/media/pubs/2005/pledgereport.pdf—last accessed February 27, 2007.

Montalbano, E. (2005) "Mozilla tests next Firefox" *PC World* (November 4, 2005).

Mordhorst, C. (1994) *Ziele und Erfolg unternehmerischer Lizenzstrategien.* Wiesbaden, Deutscher Universitätsverlag.

Mori, Y. (2005) "From practice: IP management in Japanese companies" *Management of technology and innovation in Japan*. C. Herstatt, Stockstrom, C., Tschirky, H., and Nagahira, A., Berlin, Springer: 355–385.
Morita, A. (1987) *Made in Japan—Akio Morita and Sony*. London, Collins.
Morrison, P., Roberts, J., and Midgley, D. (2004) "The nature of lead users and measurement of leading-edge status" *Research Policy* 33(2): 351–362.
Morrison, P., Roberts, J., and von Hippel, E., (2000) "Determinants of user–innovation and innovation sharing in a local market" *Management Science* 46(12): 1513–1527.
Mossberg, W. (2001) "The new Windows: Best yet, but beware—Windows XP rarely crashes, but acts as a Troyan Horse to tout Microsoft Services" *Wall Street Journal (Eastern Edition)* 238(57): B1.
Mossberger, K., Tolbert, C. and Stansbury, M. (2003) Virtual Inequality: Beyond the Digital Divide, Washington, D.D., Georgetown University Press.
Motohashi, K. (2004) *Japan's patent system and business innovation: Reassessing pro-patent policies*. Patents, Innovation and Economic Performance—2004 OECD Conference Proceedings (August 28–29, 2004), Paris, Committee for Science and Technology Policy, OECD.
Motohashi, K. (2006) *Licensing or not licensing? Empirical analysis on strategic use of patent in Japanese firms*. Tokyo, RIETI.
Motorola (1998) "Motorola announces availability of new wireless phone batteries for increased performance and safety, featuring new hologram design" Motorola, July 23, 1998.
Mowery, D. (1992) "International collaborative ventures and US firms' technology strategies" *Technology management and international business*. O. Granstrand, Hakanson, L., and Sjölander, S. Chichester, John Wiley & Sons: 209–232.
Mowery, D., and Teece, D. (1993) "Japan's growing capabilities in industrial technology" *California Management Review* 35(2): 9–34.
Mulhall, T. (1997a) "Where have all the hackers gone? Part 2—Frequency of computer related abuse: The surveys, and public perception of the problem" *Computers & Security* 16(4): 285–290.
Mulhall, T. (1997b) "Where have all the hackers gone?: Part 3—Motivation and deterrence" *Computers & Security* 16(4): 291–297.
Murmann, J., and Frenken, K. (2005) "Toward a systematic framework for research on dominant designs, technological innovations, and industrial change" *Social Science research Network*. http://ssrn.com/abstract=737063
Musker, D. (1998) "Protecting & exploiting intellectual property in electronics" *Reverse engineering*. IBC Conferences, Amsterdam, 10th June 1998.
Nagel, R. (1993) *Lead user innovationen—Entwicklungskooperationen am Beispiel der Industrie elekronischer Leiterplatten*. Wiesbaden, Deutscher Universitäts-Verlag.
Nakayama, I. (2004) *Intellectual property strategy in Japan: Towards an 'IP-based' nation*. Patents, Innovation and Economic Performance—2004 OECD Conference Proceedings 28 and 29th August, Paris, Committee for Science and Technology Policy, OECD.
Nanosys (2007) "Nanosys Overview" Nanosys. http://www.nanosysinc.com/about/history.html—last accessed February 12, 2008.
Nanotechnology Updates (2004) "Nanotechnology updates" *Nanotechnology Law & Business Journal* 1(1): Article 16.
Narin, F., and Frame, J. (1989) "The growth of Japanese science and technology" *Science* 245: 600–605.
Nash, J. (2000) "Grains of hope" *Time* 156(5): 38–46.
Nastech (2004) "Nastech acquires Cedars-Sinai Medical Center PYY patent estate for treatment of obesity" *Nastech* (May 24, 2004).
National Research Council (2004) *A patent system for the 21st century* Washington, D.C., National Academies Press.

NSB (2004) *Science and engineering indicators 2004*. Arlington, Virginia, National Science Board.

National Science Foundation (2002) *Converging technologies for improving human performance: Nanotechnology, biotechnology, information technology and cognitive science*. M. Roco and W. Bainbridge, Arlington ,Virginia, National Science Foundation.

NEC (2004) "Patent license agreement concluded between Sumitomo & NEC relating to carbon nanotube" Tokyo, NEC.

Neff, G., and Stark, D. (2004) "Permanently beta" *Society online: The internet in context*. P. A. J. Howard, S. Thousand Oaks, Sage: 173–188.

Nelson, R., Ed. (1993) *National innovation systems: A comparative analysis*. New York, Oxford University Press.

Nelson, R. (1996) *The sources of economic growth*. Cambridge, Harvard University Press.

Nelson, R. (2004) "The Market Economy and the Scientific Commons", *Research Policy* 33(3): 455-471.

Ness, E. (2005) "Stuck in the reverse" NRDC. www.nrdc.org/onearth—last accessed June 15, 2005.

Newitz, A. (2005) "Dangerous terms a user's guide to EULAs" EFF. http://www.eff.org/wp/eula.php—last accessed June 12, 2007.

Newman, E., and Golding, L. (2002) "Kabushi Kaisha Sony Computer Entertainment v Edmunds (t/a Channel Technology)" *Lovells Intellectual Property Newsletter* (February): 4–5.

Niedner, S., Hertel, G., and Hermann, S. (2000) "Motivation in free and open source projects: An internet-based survey of contributors to the Linux kernel" *Research Policy* 32: 1159–1177.

NIH (1998) *Report of the NIH Working Group on Research Tools*. National Institute of Health (June 4, 1998).

Nohr, H. (2004) "Strategie-und Geschäftsprozessorientiertes Kundenwissenmanagement" *Customer knowledge management*. H. Nohr, and Ross, A. Berlin, Logos Verlag.

Nordmann, A. (2004) *Converging technologies*. Brussels, European Commission.

Nuvolari, A. (2004) "Collective invention during the British industrial revolution; The case of the Cornish pumping engine" *Cambridge Journal of Economics* 28(3): 347–363.

Oakes, C. (1999) "DVD Hackers Hit with Lawsuit", *Wired* (December 28, 1999).

O'Mahony, S. (2003) "Guarding the commons: How community managed software projects protect their work" *Research Policy* 32(7): 1179–1198.

OECD (2001) *Biotechnology statistics in OECD member countries: Compendium of existing national statistics*. Paris, Organisation for Economic Co-operation and Development.

OECD (2004) *Patents, innovation and economic performance—OECD Conference Proceedings*. Patents, Innovation and Economic Performance—OECD Conference Proceedings August 28–29, 2004, Paris, Committee for Science and Technology Policy, OECD.

Ogawa, S. (1998) "Does sticky information affect the locus of innovation? Evidence from the Japanese convenience-store industry" *Research Policy* 26: 777–790.

Ogg, E. (2006) "Mozilla releases Firefox 2 RC1" *ZDNet News* (September 26, 2006).

Olson, E., and Bakke, G. (2001) "Implementing the lead-user method in a high-technology firm: A longitudinal study of intentions vs. actions" *Journal of Product Innovation Management* 18(2): 388–395.

Omura, J. (1997) "A patent expires, paving the road to secure e-commerce" *Network World* 14 (36): 40.

Oracle (2007) Oracle Technology Network Developer License Terms. http://www.oracle.com/technology/software/htdocs/devlic.html—last accessed May 14, 2007.

Ordon, F., and Friedt, W. (1998) *Von Mendel zum Gentransfer.* Gelsenkirchen, Th. Mann.

Ostler, U. (2003) "RFID-Special: Radio frequency identification" *www.zdnet.de* (July 7, 2003).

Oxley, J. (1999) "Institutional environment and the mechanisms of governance: The impact of intellectual property protection on the structure of interfirm alliances" *Journal of Economic Behavior and Organization* 38: 283–309.

Paarlberg, D., and Paarlberg, P. (2000) *The agricultural revolution of the 20th century.* Iowa State University Press.

Pallister, J. (2003) *A dictionary of business.* Oxford, Oxford University Press.

Pardey, P., Wright, B., and Nottenburg, K. (2001) "Are intellectual property rights stifling agricultural biotechnology in developing countries?" *International Food Policy Research Institute Annual Report 2000–2001*, Washington, D.C. IFPRI.

Pastor, M., and Sandonis, J. (2002) "Research Joint Ventures vs. Cross Licensing Agreements: An agency approach" *International Journal of Industrial Organization* 20: 215–249.

Paternoster, R. (1987) "The deterrent effect of the perceived certainty of punishment: A review of the evidence and issues" *Justice Quarterly* 4: 173–217.

Pavitt, K. (1984) "Sectoral patterns of technical change: Towards a taxonomy and a theory" *Research Policy* 13(6): 343–373.

Peace, S. (2004) "Another approach to wind" *Mechanical Engineering* (June 2004).

Pearsall, J., Ed. (1998) *The new Oxford dictionary of English.* Oxford, Oxford University Press.

Pearson, I. (2005) "Im Jahre 2030 leben wir alle in digitalen Blasen" *VDI Nachrichten.* Ipswich, England (January 21, 2005).

Perelman, M. (2002) *Steal this idea.* New York, Palgrave.

Peritz, R. (2002) "Antitrust policy and aggressive business strategy: A historical perspective on understanding commercial purposes and effects" *Journal of Public Policy & Marketing* 21(2): 237–242.

Perry, T. (2005) "DVD copy protection: Take 2" *IEEE Spectrum* (January 2005): 38–39.

Peters, T., and Waterman, R. (1988) *In search of excellence.* New York, Harper & Row.

Peterson, G., Ed. (1993) *Understanding biotechnology law—Protection, licensing and intellectual property policies.* New York, Dekker.

Petrash, G., and Bukowitz, W (1997) "Visualising, measuring and managing knowledge" *Research Technology Management* 40 (July–August): 67–74.

PEW (2004) *Genetically modified crops in the United States.* Washington, D.C. PEW.

Pharmalicensing (2006) "Report search" pharmalicensing.com/intelligence/reportsearching.php—last accessed March 3, 2007.

Phelps, C. (2005) Statement of Charles E. Phelps, Provost of Rochester, on behalf of Association of American Universities, American Council of Education, Association of American Medical Colleges and Council on Governmental Relations. *Senate Judiciary Subcommittee on Intellectual Property.*

PhRMA (2002) *Pharmaceutical industry profile 2001.* Washington, DC, Pharmaceutical Research and Manufacturers of America.

PhRMA (2007) "PhRMA mission statement" www.phrma.org/about_phrma/mission_statement/mission_statement—last accessed May 14, 2007.

Piachaud, B. (2004) *Outsourcing in the pharmaceutical industry.* New York, Palgrave Macmillan.

Pickler, N. (2002) "Mechanics struggle with diagnostics" *Associated Press* (June 24, 2002).

Pilch, H. (2004) *Why are software patents so trivial?* Patents, Innovation and Economic Performance—2004 OECD Conference Proceedings August 28–29, 2004, Paris, Committee for Science and Technology Policy, OECD.

PIPRA (2005) "Pipra-Background" Public intellectual property resource for agriculture. www.pipra.org/main/background.htm—last accessed June 12, 2007.

Pitofsky, R. (200) *Challenges of the New Economy: Issues at the Intersection of Anitrust and Intellectual Property*, Washington, D.C., American Antitrust Institute.

Pollack, P. (2006) "The next step in trusted computing makes its debut" *Ars Technica* (February 16, 2006).

Pollan, M. (1998) "Playing God in the garden" *New York Times Sunday Magazine*. New York.

Pooley, J. (2001) *Trade Secrets*, New York, Law Journal Press.

Potting Solutions (2007) "Overview of potting" http://pottingsolutions.com/my%20 site/Technology/overview_of_potting.htm—last accessed February 20, 2007.

Poynder, R. (2003a) "Enclosing the digital commons" *Information Today* 20(5): 37–38.

Poynder, R. (2003b) "Reclaiming the digital commons" *Information Today* 20(6): 33.

Prahalad, C., and Ramaswamy, V. (2000) "Co–opting customer competence" *Harvard Business Review* (January–February 2000): 79–87.

Prahalad, C., and Ramaswamy, V. (2004) *The future of competition: Co–creating unique value with customers*. Boston, MA, Harvard Business School Press.

Price, S. (1999) "Public and private plant breeding" *Nature Biotechnology* 17(10): 938.

Professional Engineering (2003) "Prius success fuels hydro vs cells debate", *Professional Engineering* (October 15, 2003).

Prügl, R., and Schreier, M. (2006) "Learning from the leading-edge customers at The Sims: Opening up the innovation process using toolkits" *R & D Management* 36(3): 237–250.

Quinn, J., and Hilmer, F. (1994) "Strategic outsourcing" *Sloan management review* 35(4): 43–55.

RAFI (1999) "USDA betrays public trust with two new terminator patents: Will USDA's biotech advisory board demand accountability?" http://www.biotech-info.net/USDA_betrays.html—last accessed June 22, 2007.

Rai, A. (1999) "Regulating scientific research: Intellectual property and the norms of science" *Northwestern University Law Review* 94(1, Fall): 77–152.

Ramírez, R. (1999) "Value co-production: Intellectual origins and implications for practice" *Strategic Management Journal* 20: 49–65.

Rammer, C., Penzkofer, H., Stephan, A. , Grenzmann, C., Heger, D., und Nagel, O. (2004) *FuE- und Innovationsverhalten von KMU und Großunternehmen unter dem Einfluss der Konjunktur, Studien zum deutschen Innovationssystem*. Nr. 22–2004. Mannheim, ZEW, IFO, DIW.

Raymond, E. (1999) *The cathedral and the bazaar: Musings on Linux and open source by an accidental revolutionary*. Sebastopol, CA, Reilly & Associates.

Razgaitis, R. (2003) *Valuation and pricing of technology-based intellectual property*. Hoboken, NJ, John Wiley & Sons.

Razgaitis, R. (2004) "US/Canadian licensing in 2003: Survey results" *LES Nouvelles* (December 2004): 139–151.

Regalado, A. (2004) "Nanotechnology patents surge as companies vie to stake claim" *Wall Street Journal (Eastern Edition)* (June 18, 2004).

Reichwald, R. and Piller, F. (2006) *Interactive Wertschöpfung*. Wiesbaden Gabler.

Reidenberg, J. (1998) "Lex informatica: The formulation of information rules through technology" *Texas Law Review* 76: 553–593.

Reiser, S., and Anbar, M., Ed. (1984) *Machine at the bedside: Strategies for using technology in patient care.* New York, Cambridge University Press.

Reiss, B. (2006) "Clinical results of CEREC inlays in a dental practice over a period of 18 years" *International Journal of Computerized Dentistry* 9(1): 11–21.

Reitzig, M. (2004) "The private values of 'thickets' and 'fences': Towards an updated picture of the use of patents across industries" *Economics of Innovation and New Technology* 13(5): 457–476.

Rescher, N. (1978) *Scientific progress: A philosophical essay on the economics of research in natural science* Oxford, Blackwell.

Response to Seattle Times (2006) "Printer fails after run-in with generic ink" www.alotofthings.com/viartshop/article_print.php?article_id=120—last accessed May 14, 2007.

Rhoades, R. (1989) "The role of farmers in the creation of agricultural technology" *Farmer First—Farmer Innovation and Agricultural Research.* R. Chambers, Pacey, A., and Thrupp, L.A. London, Intermediate Technology Publications: 3–9.

Rice, R., and Rogers, E. (1980) "Reinvention in the innovation process" *Knowledge* 1: 499–514.

Richards, P. (1985) *Indigenous agricultural revolution: Ecology and food production in West Africa.* London, Hutchinson.

Rifkin, J. (2000) *The age of access* New York, Jeremy P. Tarcher/Putnam.

Riggs, W., and von Hippel, E. (1994) "Incentives to innovate and the sources of innovation: The case of scientific instruments" *Research Policy* 23(4): 459–470.

Ring, C. Jr. (2005) "Virginia and Maryland UCITA's filling a legislative void", Journal of Internet Law 8(7): 24-31.

Rivette, K., and Kline, D. (2000) *Rembrandts in the attic: Unlocking the hidden value of patents.* Boston, Harvard Business School Press.

Roberts, E. (1988) "Managing invention and innovation: What we've learned" *Research-Technology Management* 31(Jan–Feb): 11–29.

Roberts, J. (2004) "Planters take on Monsanto over seed fees—And sometimes go to jail" *Knight Ridder Tribune Business News.* Washington (September 26, 2004): 1.

Robertson, B., and Beard, L. (2004) "Lessons learned in the deployment of alternative fuelled vehicles" *The hydrogen energy transition—Moving towards the post petroleum age in transportation.* D. Sperling, and Cannon, J., Burlington, MA, Elsevier Academic Press: 181–190.

Robinson, D. (2006) "The freedom to tinker with freedom?" http://www.freedom-to-tinker.com/?p=1046 (July 26, 2006)—last accessed June 12, 2007.

Rogers, E. (2003) *Diffusion of innovations.* New York, Free Press.

Rogers, M., and Norman, O. (2000) "Maryland follows Virginia" *Library Journal* 125(10): 7–8.

Rohner-Thielen, E. (2005) "Organic farming in Europe" *Statistics in focus: Agriculture and fisheries* (31/2005).

Roller, L.-H., Siebert, R., and Tomboy, M. (2000) *Strategic choice of partners: Research joint ventures and market power.* Discussion paper No. 2617. London, Centre for Economic Policy Research.

Rolph, D. (2004/2005) "Copyright cases" *Intellectual Property Journal* 18: 469–476.

Romer, P. (1986) "Increasing returns and long-run growth" *Journal of Political Economy* 9(5): 1002–1038.

Rosenberg, N. (1982) *Inside the black box: Technology and economics.* Cambridge, Cambridge University Press.

Rosenberg, N., and Frischtak, C. (1985) *International technology transfer.* New York, Praeger.

Rothaermel, F., and Sugiyama, S. (2001) "Virtual internet communities and commercial success: Individual and community-level theory grounded in the atypical case of TimeZone.com" *Journal of Management* 27(3): 297–312.

Rothwell, R., and Gardiner, P. (1985) "Invention, innovation, re-innovation and the role of the user: a case study" *Technovation* 3: 167–186.

Rothwell, R., and Zegveld, W. (1985) *Reindustrialization and technology.* Armonk, NY, Sharpe.

Roy, P., Fliegel, F., Kivlin, J., and Sen, L. (1968) *Agricultural innovation among Indian farmers.* Hyderabad, National Institute of Community Development.

Ruthenberg, H. (1985) *Innovation policy for small farmers in the tropics: The economics of technical innovations for agricultural development.* Oxford, Clarendon.

Safford, D. (2002) *Clarifying misinformation on TCPA.* Hawthorne, NY, IBM Research, October 2002.

Salazar, R., Louwaars, N. and Visser, B. (2003) "Issues regarding the protection of new farmers' varieties" www.kennisonline.wur.nl/NR/rdonlyres/9469C971-7C14-45A9-8AA1-D675E8AEB21A/16438/protection_varieties.doc—last accessed June 12, 2007.

Samuelson, P. (1999) "Intellectual property and the digital economy: Why the anti-circumvention regulations need revision" *Communications of the ACM (Association for Computing Machinery)* 42(9): 17–21.

Samuelson, P. (2001) "Anti-circumvention rules: Threat to science" *Science* 293 (September 14, 2001): 2028–2031.

Samuelson, P. (2002) "Reverse engineering under siege" *Communications of the ACM (Association for Computing Machinery)* 45(10): 15–20.

Samuelson, P. (2003) "DRM <and, or, vs.> the law" *Communications of the ACM (Association for Computing Machinery)* 46(4): 41–45.

Samuelson, P. (2004) "Intellectual property arbitrage: How foreign rules can affect domestic protections" *The University of Chicago Law Review* 71(1): 223–239.

Samuelson, P., and Scotchmer, S. (2002) "The law and economics of reverse engineering" *The Yale Law Journal* 100(6): 1575–1663.

Sanchez, R. (2002) "Industry standards, modular architectures, and common components: Strategic incentives for technological cooperation" F. J. Contractor, and Lorange, P., *Cooperative Strategies and Alliances.* Oxford, Pergamon: 659–687.

Saragossa, M. (2004) "EU to monitor anti–piracy efforts" www.bbc.co.uk (November 10, 2004).

Saunders, T. (2003) "Renting space on the shoulders of giants: Madey and the future of the experimental use doctrine" *Yale Law Journal* 113(1): 261–268.

Scheeres, J. (2004) "Drivers want code to their cars" *Wired* (May 31, 2004).

Scherer, F. (1977) *The economic effects of compulsory licensing.* New York, Monograph Series in Finance and Economics, Center for the Study in Financial Institutions, Graduate School of Business Administration, New York University.

Scherer, F. (1970) *Industrial market structure and economic performance.* Chicago, Rand McNally.

Schievelbein, C. (2000) "Die eigene Ernte säen" *Der Kritische Agrarbericht 2000*: 1–8.

Schiffer, M. (2003) *Draw the lightning down.* Berkeley, University of California Press.

Schimmelpfennig, D., Pray, C., and Brennan, M. (2004) "The impact of seed industry concentration on innovation: A study of US biotech market leaders" *Agricultural Economics* 30(2): 157–167.

Schlebecker, J. (1977) "Farmers and bureaucrats: Reflections on technological innovation in agriculture" *Agricultural history* 51(4): 641–655.

Schlicher, J. (1996) *Licensing intellectual property: Legal, business, and market dynamics.* New York, Wiley.

Schneier, B. (2001) *The natural laws of digital content.* Digital Libraries: Digital Asset Management, Institute of Mathematics and its Applications in Minneapolis, MN.

Schreier, M. (2005) *Wertzuwachs durch Selbstdesign—Die erhöhte Zahlungsbereitschaft von Kunden beim Einsatz von 'Toolkits for User Innovation and Design.* Wien, Deutscher Universitäts Verlag.

Schultz, E. (2002a) "The gap between cryptography and information security" *Computers & Security* 21(8): 674–676.
Schultz, E. (2002b) "Arrest for hacker following toolkit creation" *Computers & Security* 21(8): 684.
Schumpeter, J. (1934) *The theory of economic development*. Cambridge, MA, Harvard University Press.
Schumpeter, J., and Mansfield, E., Ed. (1968) "Capitalism and the process of creative destruction" in *Monopoly power and economic performance: The problem of industrial concentration*. New York, NY, Norton: 126-32.
Schweinsberg, C. (2006) "Getting mileage out of being green" *Ward's Auto World* 42(3): 34–40.
Scotchmer, S. (1991) "Standing on the shoulders of giants: Cumulative research and the patent law" *Journal of Economic Perspectives* 5: 29–41.
Scott, J. (2000) *Weapons of the weak: Everyday forms of peasant resistance*. New Haven, Yale University Press.
Sell, S. (2003) *Private power, public law: The globalization of intellectual property rights*. Cambridge, Cambridge University Press.
Senate Judiciary Committee (1998a) *WIPO copyright treatises implementation and on-line copyright infringement liability limitation*. House Report 105–551 Part 1. Washington, USA, US Senate Judiciary Committee.
Senate Judiciary Committee (1998b) *Digital Millennium Copyright Act of 1998*. House Report 105–551 Part 2. Washington, USA, Senate Judiciary Committee.
Shah, S. (2000) *Sources and patterns of innovation in a consumer products field*. Working paper, WP 4105 Sloan School of Management, Cambridge, MA, Massachusetts Institute of Technology.
Shah, S., and Tripsas, M. (2004) *When do user-innovators start firms? Towards a theory of user entrepreneurship*. Working paper, Boston, MA, Harvard Business School.
Shapiro, A. (2000) *The control revolution—How the internet is putting individuals in charge and changing the world we know*. New York, PublicAffairs.
Shapiro, C. (2001) "Navigating the patent thicket: Cross-licenses, patent pools and standard setting" *Innovation Policy and the Economy* 1: 119–50.
Shapiro, C. (2002) *Competition policy and innovation*. STI working papers 2002/11. Paris, OECD.
Shapley, D. (1978) "Electronics industry takes to potting its products for market" *Science* 202 (4370): 848–849.
Shaw, B. (1985) "The role of the interaction between the user and the manufacturer in medical equipment innovation" *R&D Management* 15(4): 283–292.
Shaw, B. (1998) "Innovation and new product development in the UK medical equipment industry", *Technoration* 4: 45-65.
Shillman, R. (2004) "A victory for real inventors" *Semiconductor International* 27(6): 127.
Shim, R. (2001) "Sony, Connectix settle PlayStation case" *ZDNet News*, March 15, 2001.
Shulman, S. (1999) *Owning the future: Inside the battles to control the new assets: Genes, software, databases, and technological know-how that make up the lifeblood of the new economy*. Boston, Houghton Mifflin.
Siegel, R. (2004) "Down but not out" *Mechanical Engineering* 126(10): 44–47.
Singel, R. (2008) "How Crypto won the DVD war" *Wired* (February 26, 2008).
Singer, M. (2005) "Coders worried by DRM chip in Apple's Intel box" *CNET News.com*, August 5, 2005.
Singh, S. (2005) "Role of the state in contract farming in Thailand" *ASEAN Economic Bulletin* 22(2): 217–228.

Sioukas, A. (1995) "User involvement for effective customization: an emperical study on voice networks", *IEEE Transactions on Engineering Management*, 42(1): 39-49.

Sirona (2004) *CEREC Hardware Compliance Requirements*, www.sirona.com/ecomaXL/index.php?site=Cenec_news_O2 – last accessed December 8, 2008.

Sirona (2005b) *Sirona Annual Report 2005*, Bensheim.

Sirona (2005a) "Sirona corporate history" Sirona. www.cereconline.com/ecomaXL/index.php?site=CEREC_history—last accessed May 16, 2008.

Sirona (2006) "CEREC 3—Operating instructions for the acquisition unit" Sirona. www.sirona.com/ecomaXL/index.php?site=SIRONA_COM_CEREC_forthe-dentist_dowloads—last accessed May 23, 2006.

Skibell, R. (2003) "Cybercrimes & misdemeanors: A reevaluation of the Computer Fraud and Abuse Act" *Berkeley Technology Law Journal* 18(3).

Slaughter, S., and Leslie, L. (1997) *Academic capitalism: Politics, policies and the entrepreneurial university.* Baltimore, MA, The Johns Hopkins University Press.

Smith Hughes, S. (2001) "Making dollars out of DNA: The first major patent in biotechnology and the commercialization of molecular biology, 1974–1980" *Isis* 92: 541–575.

Smith, J. (2005c) "Order 81" *The Ecologist* 35(1).

Smith, N. (2003) "The Shrinkwrap Snafu: Untangling the "extra element" in breach of contract claims based on shrinkwrap licenses" *Brigham Young University Law Review* 2003(4): 1373.

Solow, R. (1957) "Technical change and the aggregate production function" *The Review of Economics and Statistics* 39(3): 312–320

Somaya, D., and Teece, D. (2000) *Combining inventions in multi-invention products: Patents, organizational alternatives, and public policy.* Haas School of Business Working paper No. 99–4.

Spanbauer, S. (2003) "Product activation gains ground" *PC World* 21(6): 36.

Spann, M., Adams, M., and Souder, W. (1991) "The role of government facilities in technology transfer and local economic development" *Journal of Scientific and Industrial Research* 50: 760–770.

Spectaris (2004) *Die deutsche Medizintechnik 2004.* Berlin, Spectaris.

Sperling, L., Scheidegger, U., and Buruchara, R. (1996) *Designing seed systems with small farmers: Principles derives from bean research in the Great Lakes region of Africa.* London, Agricultural Research & Extension Network.

Spiegel, Y. (2005) *The incentive to participate in open source projects: A signalling approach.* Working paper No. 05–23, NET Institute.

Spilgies, W.-D., and Möhrle, M. (2006) "Ein Innovationsmodell für Produkt-Service-Systems" *RKW-Handbuch Führungstechnik und Organisation.* Franz, O. and Steigerwald, H. Eschborn, RKW.

Sport1 (2007) "Neuer Publikumsliebling 'Hawk Eye'" Sport1.de, http://www.sport1.de/de/sport/artikel_795795.html—last accessed February 5, 2007.

Spring, T. (2003) "Why do ink cartridges cost so much?" *PC World* (August 28, 2003).

Srinivasan, C., and Thirtle, C. (2000) "Understanding the emergence of terminator technologies" *Journal of International Development* 12: 1147–1158.

Stadtlander, C. (2005) Studie wir Sortenvielfalt im Germüsebau, Freiburg, Vorein Kultorsaat.

Stallman, R. (2002). *Free software, free society: Selected essays of Richard M. Stallman.* J. Gay. Boston, MA, GNU Press.

Staun, H. (2006) "Musik im Netz. Die Zukunft kostet nichts" *FAZ Online.* Frankfurt, (January 7, 2006).

Stedman, J. (1966) "Acquisition of patents and know-how by grant, fraud, purchase and grant-back" *University of Pittsburgh Law Review* 28: 161.
Stevans, L., and Sessions, D. (2005). "An empirical investigation into the effect of music downloading on the consumer expenditure of recorded music: A time series approach." *Journal of Consumer Policy* 28: 311–324.
Stix, G. (2005) "Best Kept Secrets", *Scientific American*, 292(1): 78-83.
Suarez-Villa, L. (2003) "The e-economy and the rise of technocapitalism: Networks, firms, and transportation" *Growth and Change* 34(4): 390–414.
Suarez-Villa, L. (2004) "Collaboration in biotechnology: How inter-firm relations strengthen research efforts in the USA" *International Journal of Technology Management* 27(5): 452–464.
Sullivan, J. (2005) *Counterattack strategies against patent trolls. Presentation to Baker & Botts' Clients in Japan.* Houston, Baker & Botts LLP.
Swann Harding, T. (1947) "Two blades of grass—A history of scientific development in the US Department of Agriculture" *The Journal of Economic History*, 7(2):254–255.
Syngenta (2004) "The story of golden rice" Syngenta. http://www.syngentafoundation.org/golden_rice.htm—last accessed February 20, 2007.
Tai, A. (2000) "Microsoft patents ASF media file format, stops reverse engineering" http://www.advogato.org/article/101.html (June 5, 2000).
Tang, P. (2005) "Digital copyright and the 'new' controversy: Is the law moulding technology and innovation?" *Research Policy* 34(6): 852–871.
Taschek, J. (2003) "Activation aggravation; It's a nasty paradox: Product activation both annoys users and brings in more money." *eWeek* 20(21): 61.
Taylor, C., and Silberston, Z. (1973) *The economic impact of the patent system: A study of the British experience.* Cambridge, Cambridge University Press.
TCG (2007) "Trusted computing group: Mobile phone work group" TCG. https://www.trustedcomputinggroup.org/groups/mobile—last accessed February 20, 2007.
Teece, D. (1977) "Technology transfer by multinational firms: The resource cost of transferring technological know-how" *The Economic Journal* 87(346): 242–261.
Telesio, P. (1977) "Foreign licensing policy in multinational enterprises" *Graduate School of Business Administration* Cambridge, MA, Harvard University.
Termini, R. (2004) *Health law: Federal regulation of drugs, biologics, medical devices, foods and dietary supplements.* Wynnewood, Pennsylvania, FORTI Publications.
The Economist (2001a) "Books and arts: The third way" *The Economist* 359 (8222): 86.
The Economist (2001b) "A dream car" *The Economist* 358(8212): 65.
The Economist (2003) "A greener bush: Energy and the environment" *The Economist* 366(8311): 13.
The Economist (2004) "An open-source shot in the arm?" *The Economist* (June 10, 2004).
The Economist (2005a) "Fingerprints for car parts" *The Economist* 377 (8456): 12.
The Economist (2005b) "Big Blue becomes bountiful" *The Economist* ( March 14, 2005): 1.
The Economist (2005c) "The rise of the creative consumer" *The Economist* (March 10, 2005).
The Economist (2006a) "Open-source business: Open, but not as usual" *The Economist* (March 16, 2006).
The Economist (2006b) "Hackers go home" *The Economist* (March 9, 2006).
The Economist (2006c) "Plugging into the future" *The Economist* (June 10, 2006).
The Royal Society & The Royal Academy of Engineering (2004) *Nanoscience and nanotechnologies: Opportunities and uncertainties* London, The Royal Society & The Royal Academy of Engineering.

Thomke, S. (2002) *Innovation at 3M corporation (A) and (B)* Harvard Business Case Studies, Cambridge, Harvard Business School Publishing.

Thomke, S., and von Hippel, E: (2002) "Customers as innovators: A new way to create value" *Harvard Business Review* 80(4): 74–81.

Tibballs, G. (1994) *The Guinness book of innovations—The 20th century from Aerosol to Zip* Enfield, Guinness Publishing.

Tietz, R., Füller, J., and Herstatt, C. (2006) "Signalling—An innovative approach to identify lead users in online communities" *Customer Interaction and Customer Integration.* T. Blecker, and Friedrich, G. Berlin, GITO-Verlag: 453–467.

Tietz, R., Herstatt, C., Morrison, P. and Lüthje, C. (2005) "The process of user-innovation: A case study in a consumer goods setting" *International Journal of Product Development* 2(4): 321—338.

Tittle, C. (1969) "Crime rates and legal sanctions" *Social Problems* 16: 409–423.

Tiwari, R., and Buse, S. (2007) *The mobile commerce prospects: A strategic analysis of opportunities in the German banking sector.* Hamburg, Hamburg University Press.

Trilateral (1996) *Trilateral statistical report 1996.* Trilateral (EPO, JPO and USPTO).

Trilateral (2000) *Trilateral statistical report 2000.* Trilateral (EPO, JPO and USPTO).

Trilateral (2005) *Trilateral statistical report 2005.* Trilateral (EPO, JPO and USPTO).

Trott, P. (1998) *Innovation management & new product development.* London, Financial Times.

Tsilas, N. (2005) "Recently proposed definitions of 'open standards' could impede innovation, interoperability, and government procurement options" *LES Nouvelles* (September 2005): 105–111.

Tuomi, I. (2006) *Networks of innovation: Change and meaning in the age of the internet.* Oxford, Oxford University Press.

Tuxill, J. (1999) *Nature's cornucopia: Our stake in plant diversity.* Washington, D.C., Worldwatch Institute.

TWST (The Wall Street Transcript Corporation) (2004) "Roundtable forum: Storage hardware & software" *Wall Street Transcript.* New York: 1.

Udell, J. (2003) "Lizard brain surgery" *InfoWorld* 25(45): 48.

Ughanwa, D., and Baker, M. (1989) *The role of design in international competitiveness.* London, Routledge.

UNDP (2000) *Human development report 2000: Human rights and human Development.* New York, United Nations Development Programme.

UNEP Convention on Biological Diversity (2003) *Report of the ad hoc technical expert group meeting on the potential impacts of genetic use restriction technologies on smallholder farmers, indigenous and local communities and farmers' rights.* UNEP/CBD/SBSTTA/9/INF/6. Montreal, United Nations Environmental Programme.

UNESCO (2005) *EFA global monitoring report 2005.* Paris, UNESCO.

UNFPA (2004) *UNFPA state of world population 2004.* United Nations Population Fund.

UNFPA (2006) *UNFPA state of world population 2006* United Nations Population Fund.

UPOV (2006) *Plant variety protection statistics for the period 2001–2005* Geneva, International Union for the Protection of new varieties of Plants.

Urban, G., and von Hippel, E. (1988) "Lead user analyses for the development of new industrial products" *Management Science* 34(5): 569–582.

U.S. Copyright Office (2001) *DMCA Section 104 Report*, U.S. Copyright Office, Washington, D.C.

USDA (2001) *Why USDA's technology protection system (aka "Terminator") benefits Agriculture?* Washington, D.C., Agricultural Research Service, United States Department of Agriculture.

USDJ (1998) "Juvenile computer hacker cuts off FAA tower at regional airport" US Department of Justice, March 18. www.cybercrime.gov/juvenilepld.htm—last accessed February 14, 2007.
USPTO (2008) "Calendar year 2006 patent counts by patent counts, by patent type and by state and country of origin" USPTO. ftp://ftp.uspto.gov/pub/taf/st_co_06.htm—last accessed April 17, 2008.
USPTO (2008b) "US patent statistics chart calendar years 1963–2007" USPTO. http://www.uspto.gov/web/offices/ac/ido/oeip/taf/us_stat.htm—last accessed April 17, 2008.
USPTO (2006) "US patent activity calendar years 1790 to the present" USPTO. http://www.uspto.gov/go/taf/h_counts.pdf—last accessed March 12, 2008.
Utterback, J. (1994) *Mastering the dynamics of innovation*. Boston, MA, Harvard Business School Press.
Valenti, J. (2000) "If you can't protect what's yours, you own nothing" *L.A. Times, Opinion Piece* (January 30, 2000).
Van Dijk, T. (2000) "License contracts, future exchange clauses and technological competition" *European Economic Review* 44(8): 1431–1448.
Varchaver, N. (2001) "The patent king" *Fortune* 143(10): 202–212.
Varian, H. (2002) "New chips can keep a tight rein on consumers" *New York Times* New York (July 4, 2002).
Vaughan-Nichols, S. (2004) "Author of Linux Patent Study says Ballmer got it wrong" *eWeek* (November 19, 2004).
Vernooy, R. (2003) *Seeds that give: Participatory plant breeding*. Ottawa, International Development Research Centre.
Veryzer Jr., R. (1998a) "Discontinuous innovation and the new product development process" *Journal of Product Innovation Management* 15(4): 304–321.
Veryzer Jr., R. (1998b) "Key factors affecting customer evaluation of discontinuous new products" *Journal of Product Innovation Management* 15(2): 136–150.
Vestas (2005) *Vestas annual report 2005*. Randers Vestas.
Veverka, M. (2003) "The next big thing is really amazingly small" *Barron's* 83(29): T2.
Vickery, G. (1988) "A survey of international technology licensing" *STI Review* 4: 7–49.
Virmani, S. (1998) *Advances in Hybrid Rice Technology*. Los Baños, IRRI.
Virmani, S., Mao, C., Toledo, R., Hossain, M., and Janaiah, A. (2002) "Hybrid rice seed production technology and its impact on seed industries and rural employment opportunities in Asia" Taipei, Food and Fertilizer Technology Center, *Technical bulletin*: 156.
Visionary (2006) "Terms and conditions" Visionary AV Solutions Ltd. http://www.plasmas-direct.co.uk/standard/terms_and_conditions.htm—last accessed February 15, 2007.
Volling, A. (2006) "Der Monsanto-Vertrag -Lupe lohnt sich" *Bauernstimme* June 2006(6).
von dem Bussche, P. (2004) "Globale Verantwortung" *ISF Congress*. Berlin (May 24, 2004).
von Hippel, E. (1976) "The dominant role of users in the scientific instrument innovation process" *Research Policy* 5: 212–239.
von Hippel, E. (1977) "The dominant role of the user in semiconductor and electronic subassembly process innovation" *IEEE Transactions on Engineering Management* 24(2): 60–71.
von Hippel, E. (1981) "Appropriability of innovation benefit as a predictor of the source of innovation" *Research Policy* 11: 95–115.
von Hippel, E. (1985) *Testing the correlation between the functional locus of innovation and appropriable innovation benefit*. Cambridge, MA, MIT.
von Hippel, E. (1988) *The sources of innovation*. Oxford, Oxford University Press.

von Hippel, E. (1994) "'Sticky information' and the locus of problem solving: Implications for innovation" *Management Science* 40(4): 429–439.
von Hippel, E. (2001) *Open source software: Innovation by and for users—No manufacturer required!* Cambridge, MA, MIT.
von Hippel, E. (2005a) *Democratizing innovation.* Cambridge, MA, MIT Press.
von Hippel, E. (2005b) "Democratizing innovation: The evolving phenomenon of user-innovation" *Journal for Betriebswirtschaft* 55(1): 63–78.
von Hippel, E., and Finkelstein, S. (1979) "Analysis of innovation in automated clinical chemistry analyzers" *Science and Public Policy* 6: 24–37.
von Hippel, E., and Katz, R. (2002) "Shifting innovation to users via toolkits" *Management Science* 48(7): 821–833.
von Hippel, E., and von Krogh, G. (2006) "Free revealing and the private-collective model for innovation incentives" *R&D Management* 36(3): 295–306.
von Hippel, E., Thomke, S., and Sonnack, M. (2000) "Creating breakthroughs at 3M" *Health Forum Journal* 43(4): 20–28.
Von Weizsäcker, C. (1980) "The cost of substitution" *Econometrica* 52: 1085–1116.
Vonortas, N., and Kim, Y. (2004) *Technology licensing.* Patents, Innovation and Economic Performance—2004 OECD Conference Proceedings (August 28–29, 2004, Paris, Committee for Science and Technology Policy, OECD.
Voss, C. (1985) "The role of users in the development of applications software" *Journal of Product Innovation Management* 2: 113–121.
Wagner, H. (2003) "Legislators, academics propose new anti-piracy alternatives" *Video Store Magazine* 25(12): 14.
Wagner, S., and McDermott, J. (2006) "Unternehmen Gott" *Focus* 42(16): 166–171.
Wagstaff, J. (2002) "How to squeeze more juice from pricey printer cartridges" *Wall Street Journal Online*, (August 6, 2002).
Walsh, J., Arora, A., and Cohen, W. (2003) "Effects on research tool patents and licensing on biomedical innovation" *Patents in the knowledge-based economy.* Board on Science, Technology and Economic Policy, Washington, D.C., National Academic Press: 285-340.
Washburn, J. (2005) *University, Inc.—The corporate corruption of higher education.* New York, Basic Books.
Wearden, G. (2001) "Robotics enthusiast forced to pull Aibo–altering code" *ZDNet UK* (November 1, 2001).
Weber, S. (2004) *The success of open source.* Cambridge, MA, Harvard University Press.
Werker, C. and Athreye, S. (2004) "Marshall's disciples: knowledge and innovation driving regional economic development and growth", *Journal of Evolutionary Economics*, 14(5): 505-523.
Weihermüller, M. (1982) *Die Lizenzvergabe im internationalen marketing: Entscheidungsgrundlagen Gestaltungsbereiche.* München, Florentz Verlag.
Weisbrod, B. (1994) "The nature of technological change; Incentives matter!" *The changing economics of medical technology.* A. Gelijns, and Dawkins, H. Washington, D.C, National Academy Press: 8–48.
Wellen, A., and Reiter, L. (1998) "Are hackers criminals?" *TechTV vault*. http://www.g4tv.com/techtvvault/features/1161/Are_Hackers_Criminals.html—last accessed June 13, 2007.
Westphal, L., Kim, L., and Dahlman, C. (1985) "Reflections on The Republic of Korea's acquisition of technological capability" *International technology transfer.* N. Rosenberg, and Frischtak, C. New York, Praeger: 167–221.
Whitmore, E. (1997) *Product development planning for health care products regulated by the FDA.* Milwaukee, Wisconsin, ASQC Quality Press.
Wible, B. (2003) "A site where hackers are welcome: Using-hack in contests to shape preferences and deter computer crime" *The Yale Law Journal* 112(6): 1577–1623.

Wiedhahn, K. (2006) "20 Years CEREC" *International Journal of Computerized Dentistry* 9(1).

Wikipedia (2007) "Source line of code" Wikipedia. http://en.wikipedia.org/wiki/Source_lines_of_code—last accessed February 19, 2007.

Wilke, J., and Bandler, J. (2001) "Kodak tangles with Microsoft over Win XP" *Wall Street Journal* (July 2, 2001).

Willer, H., and Yussefi, M. (2006) *The world of organic agriculture. statistics and emerging trends 2006* Bonn, International Federation of Organic Agriculture Movements (IFOAM).

Williams, A. (2005) *The patent pxplosion.* New Paradigm Learning Corporation.

Wilson, J. (2004) "Can a copyright holder prevent reverse engineering? The federal circuit court holds that the Federal Copyright Act does not preempt 'no reverse engineering' clauses" *Computer Law Review and Technology Journal* 8: 467–475.

Wilson, R. (1977) "International licensing of technology: Empirical evidence" *Research Policy* 6(2): 114–126.

Winston, K. (2005) "The internal morality of Chinese legalism" *Singapore Journal of Legal Studies*, December: 31.

Winter, S., and Szulanski, G. (2001) Replication as strategy" *Organization Science* 12(6): 730–743.

Witcombe, J. (2005) "Participatory varietal selection and participatory plant breeding: The last 10 years" *Participatory research and development for sustainable agriculture and natural resource management: A sourcebook—Volume 1: Understanding participatory research and development.* J. Gonsalves, Becker, T., Braun, A., Campilan, D., Chavez, H., Fajber, E., Kapiriri, M., Rivaca-Caminade, J. and Vernooy, R. Canada, The International Development Research Centre (IDRC).

Witcombe, J., Joshi, A., Joshi, K. and Sthapit, B. (1996) "Farmer participatory crop improvement: Varietal selection and breeding methods and their impact on biodiversity" *Experimental Agriculture* 32(4): 445–460.

Woodward, W., Jr. (2002) "UCITA's imperialism" *The Licensing Journal* (March 2002): 1–8.

Wragge & Co. (2004) "UK regulations: Controversial EU copyright directive implemented" *EIU ViewsWire.* New York (February 9, 2004).

Wüstenhagen, R. (2003) "Sustainability and competitiveness in the renewable energy sector: The case of Vestas Wind Systems" *Greener Management International* 44: 105–116.

Yanagawa, N., and Wada, T. (2000) *Post-contracting innovations and contingent payment scheme in patent licensing contracts.* 4th ISNIE Conference "The Economics of Institutions in the New Millennium" September 22–24, 2000, Tübingen, Germany, ISNIE.

Yang, G., and Maskus, K. (2001) "Licensing and intellectual property rights: An empirical investigation" *Weltwirtschaftliches Archiv* 137: 58–79.

Zittrain, J. (2000) "What the publisher can teach the patient: Property and privacy in an era of trusted privication" *Stanford Law Review* 52: 1201.

Zittrain, J. (2005) "In praise of uncertainty" *Harvard Business Review* 83(5): 18–20.

Zyvex (2007) Atomically precise manufacturing. www.zyvex.com/Research/activities—last accessed February 20, 2007.

# Index